Sydney John Hickson

The Fauna of the Deep Sea

Sydney John Hickson

The Fauna of the Deep Sea

ISBN/EAN: 9783337320836

Printed in Europe, USA, Canada, Australia, Japan

Cover: Foto ©berggeist007 / pixelio.de

More available books at **www.hansebooks.com**

THE
FAUNA OF THE DEEP SEA

BY

SYDNEY J. HICKSON, M.A. (CANTAB. ET OXON.)

D.SC. (LOND.), FELLOW OF DOWNING COLLEGE, CAMBRIDGE

WITH TWENTY-THREE ILLUSTRATIONS

LONDON
KEGAN PAUL, TRENCH, TRÜBNER & CO. LTD.
PATERNOSTER HOUSE, CHARING CROSS ROAD
1891

PREFACE

THE time may come when there will be no portion of the earth's surface that has not been surveyed and explored by man.

The work of enterprising travellers has now been carried on within a measurable distance of the North Pole; the highest mountain ranges are gradually succumbing to the geological surveyor; the heart of Africa is giving up to us its secrets and its treasures, and plans of all the desert places of the earth are being made and tabulated.

The bottom of the deep sea was until quite recently one of these terræ incognitæ. It was regarded by most persons, when it entered into their minds to consider it at all, as one of those regions about which we do not know anything, never shall know anything, and do not want to know anything.

But the men of science fifty years ago were not disposed to take this view of the matter. Pushing their inquiries as to the character of the sea-fauna into deeper and deeper water, they at length demanded information as to the existence of forms of animal life in the greatest depths. Unable themselves to bear the heavy expenses involved in such an investigation, they sought for and obtained the assistance of the Government, in the form of national ships, for the work, and then our knowledge of the depths of the great ocean may be said to have commenced.

We know a good deal now, and in the course of time we may know a great deal more, about this interesting region; but it is not one which, in our generation at any rate, any human being will ever visit.

We may be able to plant the Union Jack on the summit of Mount Everest, we may drag our sledges to the South Pole, and we may, some day, be able to travel with ease and safety in the Great Sahara; but we cannot conceive that it will ever be possible for us to invent a diving-bell that will take a party of explorers to a depth of three and a half miles of water. We may complete our survey of the ocean beds, we may analyse the bottom muds and name and classify the animals that compose their fauna, but there are many things that must remain merely matters of conjecture. We shall never know, for example, with any degree of certainty, how *Bathypterois* uses its long feeler-like pectoral fins, nor the meaning of the fierce armature of *Lithodes ferox*; why the deep-sea Crustacea are so uniformly coloured red, or the intensity of the phosphorescent light emitted by the Alcyonaria and Echinoderms. These and many others are and must remain among the mysteries of the abyss.

Our present-day knowledge of the physical conditions of the bottom of the deep sea and the animals that dwell there is by no means inconsiderable.

It may be found in the reports of the scientific expeditions fitted out by the English, French, German, Italian, Norwegian, and American Governments, in numerous volumes devoted to this kind of work, and in memoirs and notes scattered through the English and foreign scientific journals.

It is the object of this little book to bring together

in a small compass some of the more important facts and considerations that may be found in this great mass of literature, and to present them in such a form that they may be of interest to those who do not possess a specialist's knowledge of genera and species.

When it was found that animals can and do live even at the greatest depths of the ocean, the interest of naturalists was concentrated on the solution of the following problems. Firstly, do the animals constituting the fauna of the abyss exhibit any striking and constant modification in correlation with the physical conditions of their strange habitat? And, secondly, from what source was the fauna of the abyss derived? Was it derived from the shallow shore waters, or from the surface of the sea? Is it of very ancient origin, or the result of, comparatively speaking, recent immigrations?

These questions cannot be answered in a few lines. Any views that may be put forward regarding them require the support of a vast array of facts and figures; but as the limits of this little book would not permit of my giving these, I have endeavoured to select a few only of those which bear most directly upon the points at issue.

To overburden my work with the names of genera or the lists of species would not, it seemed to me, either clear the issues or interest the general reader. These may be found in the 'Challenger' monographs, and other books dealing with the subject.

Those who wish to pursue the subject further will find in the 'Voyages of the "Blake,"' by Alexander Agassiz, an excellent and elaborate discussion of deep-sea problems, and numerous illustrations of some of the most interesting forms of abysmal life.

In Volume XXIII. of the 'Bulletin of Comparative Zoology' the same author gives a most interesting account of the deep-sea work that has recently been done by the 'Albatross' expedition.

Filhol's 'La Vie au Fond des Mers' is also a book that contains a great deal of new and interesting matter, together with some excellent coloured plates of deep-sea animals.

SYDNEY J. HICKSON.

DOWNING COLLEGE, CAMBRIDGE:
September, 1893.

CONTENTS

CHAPTER		PAGE
I.	A Short History of the Investigations	1
II.	The Physical Conditions of the Abyss	17
III.	The Relations of the Abysmal Zone and the Origin of its Fauna	45
IV.	The Characters of the Deep-sea Fauna	59
V.	The Protozoa, Cœlentera, and Echinoderma of the Deep Sea	86
VI.	The Vermes and Mollusca of the Deep Sea	109
VII.	The Arthropoda of the Deep Sea	123
VIII.	The Fish of the Deep Sea	148
Index		167

LIST OF ILLUSTRATIONS

STOMIAS BOA. AFTER FILHOL, 'LA VIE AU FOND
DES MERS'. . . . *Frontispiece*

FIG.		PAGE
1	DIAGRAM ILLUSTRATING THE PASSAGE OF AN OCEAN CURRENT ACROSS A BARRIER . .	32
2	*Sicyonis crassa*. AFTER HERTWIG, '"CHALLENGER" REPORTS'	36
3	GLOBIGERINA OOZE. AFTER AGASSIZ, 'VOYAGES OF THE "BLAKE"'	38
4	SECTION THROUGH THE EYE OF *Serolis schythei*. AFTER BEDDARD, '"CHALLENGER" REPORTS'	74
5	SECTION THROUGH THE EYE OF *Serolis bromleyana*. AFTER BEDDARD, '"CHALLENGER" REPORTS'	74
6	*Opostomias micripnus*. AFTER GÜNTHER, '"CHALLENGER" REPORTS'	78
7	HEAD OF *Pachystomias microdon*. AFTER VON LENDENFELD, '"CHALLENGER" REPORTS' .	79
8	SECTION THROUGH THE ANTERIOR SUB-ORBITAL PHOSPHORESCENT ORGAN OF *Pachystomias microdon*. AFTER VON LENDENFELD, '"CHALLENGER" REPORTS' . .	80

FIG.		PAGE
9	*Challengeria Murrayi.* AFTER HAECKEL, '"CHALLENGER" REPORTS'	90
10	*Umbellula Güntheri.* AFTER AGASSIZ, 'VOYAGES OF THE "BLAKE"'	97
11	*Rhizocrinus lofotensis.* AFTER CARPENTER, '"CHALLENGER" REPORTS' . .	100
12	*Rhabdopleura normani.* AFTER LANKESTER, 'CONTRIBUTIONS TO OUR KNOWLEDGE OF RHABDOPLEURA AND AMPHIOXUS'	112
13	A SINGLE POLYPIDE OF *Rhabdopleura normani.* AFTER LANKESTER, tom. cit. . . .	114
14	*Bathyteuthis abyssicola.* AFTER HOYLE, '"CHALLENGER" REPORTS'	121
15	*Bathynomus giganteus.* AFTER FILHOL, tom. cit.	131
16	*Euphausia latifrons.* AFTER SARS, '"CHALLENGER" REPORTS'	134
17	*Bentheuphausia amblyops.* AFTER SARS, "CHALLENGER" REPORTS'	134
18	*Polycheles baccata.* AFTER SPENCE BATE, '"CHALLENGER" REPORTS'	136
19	*Collosendeis arcuatus.* AFTER FILHOL, tom. cit.	141
20	*Hypobythius calycodes.* AFTER MOSELEY, '"CHALLENGER" REPORTS' . . .	145
21	*Melanocetus Murrayi.* AFTER GÜNTHER, '"CHALLENGER" REPORTS'	156
22	*Saccopharynx ampullaceus.* AFTER GÜNTHER, '"CHALLENGER" REPORTS'	164

THE FAUNA OF THE DEEP SEA

CHAPTER I

A SHORT HISTORY OF THE INVESTIGATIONS

OUR knowledge of the natural history of the deep seas may be said to have commenced not more than fifty years ago. There are, it is true, a few fragments of evidence of a fauna existing in depths of more than a hundred fathoms to be found in the writings of the earlier navigators, but the methods of deep-sea investigation were so imperfect in those days that naturalists were disposed to believe that in the abysses of the great oceans life was practically non-existent.

Even Edward Forbes just before his death wrote of an abyss 'where life is either extinguished or exhibits but a few sparks to mark its lingering presence,' but in justice to the distinguished naturalist it should be remarked that he adds, 'Its confines are yet undetermined, and it is in the exploration of this vast deep-sea region that the finest field for submarine discovery yet remains.'

Forbes was only expressing the general opinion of naturalists of his time when he refers with evident hesitation to the existence of an azoic region. His own dredging excursions in depths of over one hundred fathoms proved the existence of many peculiar species that were previously unknown to science. 'They were like,' he says, 'the few stray bodies of strange red men, which tradition reports to have been washed on the shores of the Old World before the discovery of the New, and which served to indicate the existence of unexplored realms inhabited by unknown races, but not to supply information about their character, habits, and extent.'

In the absence of any systematic investigation of the bottom of the deep sea, previous to Forbes's time the only information of deep-sea animals was due to the accidental entanglement of certain forms in sounding lines, or to the minute worms that were found in the mud adhering to the lead.

As far back as 1753, Ellis described an Alcyonarian that was brought up by a sounding line from a depth of 236 fathoms within eleven degrees of the North Pole by a certain Captain Adriaanz of the 'Britannia.' The specimen was evidently an *Umbellula*, and it is stated that the arms (i.e. Polyps) were of a bright yellow colour and fully expanded when first brought on deck.

In 1819 Sir John Ross published an account of his soundings in Baffin's Bay, and mentions the existence of certain worms in the mud brought from a depth of 1,000 fathoms, and a fine Caput Medusæ (Astrophyton) entangled on the sounding line at a depth of 800 fathoms.

In the narrative of the voyage of the 'Erebus' and 'Terror,' published in 1847, Sir James Ross calls attention to the existence of a deep-sea fauna, and makes some remarks on the subject that in the light of modern knowledge are of extreme interest. 'I have no doubt,' he says, 'that from however great a depth we may be enabled to bring up the mud and stones of the ocean, we shall find them teeming with animal life.' This firm belief in the existence of an abysmal fauna was not, as it might appear from the immediate context of the passage I have quoted, simply an unfounded speculation on his part, but was evidently the result of a careful and deliberate chain of reasoning, as may be seen from the following passage that occurs in another part of the same book :—' It is well known that marine animals are more susceptible of change of temperature than land animals; indeed they may be isothermally arranged with great accuracy. It will, however, be difficult to get naturalists to believe that these fragile creatures could possibly exist at the depth of nearly 2,000 fathoms below the surface; yet as we know they can bear the pressure of 1,000 fathoms, why may they not of two? We also know that several of the same species of creatures inhabit the Arctic that we have fished up from great depths in the Antarctic seas. The only way they could get from one pole to the other must have been through the tropics; but the temperature of the sea in those regions is such that they could not exist in it, unless at a depth of nearly 2,000 fathoms. At that depth they might pass from the Arctic to the Antarctic Ocean without a variation of five degrees of temperature; whilst any land animal, at the most favourable season, must experience a difference of fifty degrees, and, if in the winter, no less than 150 degrees of Fahrenheit's thermometer—a sufficient reason why there are neither quadrupeds, nor birds, nor land insects common to both regions.'

In the year 1845, Goodsir succeeded in obtaining a good haul in Davis Straits, at a depth of 300 fathoms. It included Mollusca, Crustacea, Asterids, Spatangi, and Corallines.

In 1848, Lieutenant Spratt read a paper at the meeting of the British Association at Swansea, on the influence of temperature upon the distribution of the fauna in the Ægean seas, and at the close of this paper we find the following passage, confirming in a remarkable way the work of previous investigators in the same field. He says: 'The greatest depth at which I have procured animal life is 390 fathoms, but I believe that it exists much lower, although the general character of the Ægean is to limit it to 300 fathoms; but as in the deserts we have an oasis, so in the great depths of 300, 400, and perhaps 500 fathoms we may have an oasis of animal life amidst the barren fields of yellow clay dependent upon favourable and perhaps accidental conditions, such as the growth of nullipores, thus producing spots favourable for the existence and growth of animal life.'

The next important discovery was that of the now famous Globigerina mud by Lieutenants Craven and Maffit, of the American Coast survey, in 1853, by the help of the sounding machine invented by

Brooke. This was reported upon by Professor Bailey.

Further light was thrown upon the deep-sea fauna by Dr. Wallich in 1860, on board H.M.S. 'Bulldog,' by the collection of thirteen star-fish living at a depth of 1,260 fathoms.

Previous to this Torell, during two excursions to the Northern seas, had proved the existence of an extensive marine fauna in 300 fathoms, and had brought up with the 'Bulldog' machine many forms of marine invertebrates from depths of over 1,000 fathoms; but it was not until 1863, when Professor Lovén read a report upon Torell's collections, that these interesting and valuable investigations became known to naturalists.

Nor must mention be omitted of the remarkable investigations of Sars and his son, the pioneers of deep-sea zoology on the coasts of Norway, who, by laborious work commenced in 1849, failed altogether to find any region in the deep water where animal life was non-existent, and indeed were the first to predict an extensive abysmal fauna all over the floor of the great oceans. One of the many remarkable discoveries made by Sars was *Rhizocrinus*, a stalked Crinoid.

A SHORT HISTORY OF THE INVESTIGATIONS 7

Ever since that time the Norwegians and the Swedes have been most energetic in their investigations, and the publications of the results of the Norske Nord-havns expeditions are regarded by all naturalists as among the most valuable contributions to our knowledge of the deep-sea fauna.

Notwithstanding the previous discovery of many animals that undoubtedly came from the abysmal depths of the ocean, there were still some persons who found a difficulty in believing that animal life could possibly exist under the enormous pressure of these great depths. It was considered to be more probable that these animals were caught by the dredge or sounding lines in their ascent or descent; and that they were merely the representatives of a floating fauna living a few fathoms below the surface.

The first direct proof of the existence of an invertebrate fauna in deep seas was found by the expedition that was sent to repair the submarine cable of the Mediterranean Telegraph Company. The cable had broken in deep water, and a ship was sent out to examine and repair the damage. When the broken cable was brought on deck, it bore several forms of animal life that must have become attached to it and lived at the bottom of the sea in water extending down to a depth of 1,200 fathoms. Among other forms a *Caryophyllia* was found attached to the cable at 1,100 fathoms, an oyster (*Ostrea cochlear*), two species of Pecten, two gasteropods, and several worms.

The discoveries that had been made indicating the existence of a deep-sea fauna led to the commission of H.M. ships 'Lightning' and 'Porcupine,' and the systematic investigation that was made by the naturalists on these vessels brought home to the minds of naturalists the fact that there is not only an abysmal fauna, but that in places this deep-sea fauna is very rich and extensive. The 'Lightning' was despatched in the spring of 1868 and carried on its investigations in the neighbourhood of the Faeroe Islands, but the vessel was not suitable for the purpose and met with bad weather. The results, however, were of extreme importance; for, besides solving many important points concerning the distribution of ocean temperature, 'it had been shown beyond question that animal life is both varied and abundant at depths in the ocean down to 650 fathoms at least, notwithstanding the extraordinary conditions to which animals are there exposed.'

Among the remarkable animals dredged by the

'Lightning' were the curious Echinoderm, *Brisinga coronata*, previously discovered by Sars, and the Hexactinellid sponges, *Holtenia* and *Hyalonema*, the Crinoids *Rhizocrinus* and *Antedon celticus*, and the Pennatulid *Bathyptilum Carpenteri*, not to mention numerous Foraminifera new to science.

In the spring of the following year, 1869, the Lords Commissioners of the Admiralty despatched the surveying vessel 'Porcupine' to carry on the work commenced by the 'Lightning.'

The first cruise was on the west coast of Ireland, the second cruise to the Bay of Biscay, where dredging was satisfactorily carried on to a depth of 2,435 fathoms, and the third in the Channel between Faeroe and Scotland.

The dredging in 2,435 fathoms was quite successful, and the dredge contained several Mollusca, including new species of *Dentalium*, *Pecten*, *Dacrydium*, &c., numerous Crustacea and a few Annelids and Gephyrea, besides Echinoderma and Protozoa. A satisfactory dredging was also made in 1,207 fathoms.

The third cruise was also successful and brought many new species to light, including the *Porocidaris purpurata*, and a remarkable heart urchin, *Pourtalesia Jeffreysi*.

Concerning Pourtalesia Sir Wyville Thomson says:—

'The remarkable point is that, while we had so far as we were aware no living representative of this peculiar arrangement of what is called "disjunct" ambulacra, we have long been acquainted with a fossil family—the Dysasteridæ—possessing this character. Many species of the genera Dysaster, Collyrites, &c., are found from the lower oolite to the white chalk, but there the family had previously been supposed to have become extinct.'

The discovery of two new Crinoids led to the anticipation that the Crinoidea, the remarkable group of Echinoderma, supposed at the time to be on the verge of extinction, probably form rather an important element in the abysmal fauna.

One of the most interesting results was the discovery of three genera in deep water, *Calveria*, *Neolampas* and *Pourtalesia*, almost immediately after they were discovered by Pourtales in deep water on the coasts of Florida, showing thus a wide lateral distribution and suggesting a vast abysmal fauna.

A year before the 'Lightning' was despatched, Count Pourtales had commenced a series of investigations of the deep-sea fauna off the coast of Florida. The first expedition started in 1867 from Key West for the purpose of taking some dredgings between that port and Havana. Unfortunately yellow fever broke out on board soon after they started, and only a few dredgings were taken. However, the results obtained were of such importance that they encouraged Pourtales to undertake another expedition and enabled him to say very positively ' that animal life exists at great depths, in as great a diversity and as great an abundance as in shallow water.'

In the following years, 1868 and 1869, the expeditions were more successful, and many important new forms were found in water down to 500 fathoms. Perhaps the most interesting result obtained was the discovery of *Bourguetticrinus* of D'Orbigny ; it may even be the species named by him which occurs fossil in a recent formation in Guadeloupe.

By this time the interest of scientific men was thoroughly excited over the many problems connected with this new field of work. The prospect of obtaining a large number of new and extremely curious animals, the faint hope that living Trilobites, Cystids, and other extinct forms might be discovered, and lastly the desire to handle and investigate great masses of pure protoplasm in the form of the famous but unfortunately non-existent Bathybius, induced some men of wealth and leisure to spend their time in deep-sea dredging, and stimulated the governments of some civilised countries to lend their aid in the support of expeditions for the deep-sea survey.

Mr. Marshall Hall's yacht, the 'Norma,' was employed for some time in this work, and an extensive collection of deep-sea animals was made. About the same time Professor L. Agassiz was busy on board the American ship, the 'Hassler,' in continuing the work of Count Pourtales, and later on the Germans fitted out the 'Gazelle,' and the French the still more famous 'Travailleur' and 'Talisman' expeditions. Nor must we omit to mention in this connection the cruise of the Italian vessel, the 'Vittor Pessani,' nor those of the British surveying vessels, the 'Knight Errant' and the 'Triton,' and the American vessels, 'The Blake' and the 'Fish Hawk.'

But of all these expeditions, by far the most complete in all the details of equipment, and the arrangements made for the publication of the results, was the expedition fitted out in 1873 by the British Government. The voyage of H.M.S. 'Challenger' is so familiar to all who take an interest in the progress of scientific discovery, that it is not necessary to do more than make a passing mention of it in this place. The excellent books that were written by Wyville Thomson, by Moseley, and by other members of the staff, have made the general reader familiar with the narrative of that remarkable cruise and the most striking of the many scientific discoveries that were made; while the numerous large monographs that have been published during the past fourteen years give opportunities to the naturalist of obtaining all the requisite information concerning the detailed results of the expedition.

The expenditure of the large sum of money upon this expedition and the publication of its reports has been abundantly justified. The information obtained by the 'Challenger' will be for many years to come the nucleus of our knowledge of the deep-sea fauna, the centre around which all new facts will cluster, and the guide for further investigations.

To say that the 'Challenger' accomplished all that was expected or required would be to over-estimate the value of this great expedition, but nevertheless it is difficult for us, even now, thoroughly to grasp the importance of the results obtained or to analyse and classify the numerous and very remarkable facts that were gained during her four years' cruise.

It is, of course, impossible, in a few lines, to give a summary of the more important of the Natural History results of the 'Challenger' expedition. Besides proving the existence of a fauna in the sea at all depths and in all regions, the expedition further proved that the abysmal fauna, taken as a whole, does not possess characters similar to those of the fauna of any of the secondary or even tertiary rocks. A few forms, it is true, known to us up to that time only as fossils, were found to be still living in the great depths, but a large majority of the animals of these regions were found to be new and specially modified forms of the families and genera inhabiting shallow waters of modern times. No Trilobites, no Blastoids, no Cystoids, no new Ganoids, and scarcely any deep-sea Elasmobranchs were brought to light, but the fauna was found to consist mainly of Teleosteans, Crustacea, Cœlentera, and other creatures unlike anything known to have existed in Palæozoic times, specially modified in structure for their life in the great depths of the ocean.

In 1876 the s.s. 'Vöringin' was chartered by the Norwegian Government and was dispatched to investigate the tract of ocean lying between Norway, the Faeroe islands, Jan Mayen, and Spitzbergen. The investigations extended over three years, the vessel returning to Bergen in the winter months.

The civilian staff of the 'Vöringin' included Professors H. Mohn, Danielssen, and G. O. Sars, and the expedition was successful in obtaining a large number of animals from deep water by means of the dredge and tangles and by the trawl.

The results of this expedition have been published in a series of large quarto volumes under the general title of the Norske Nord-havns Expedition.

The most interesting forms brought to light by the Norwegians are the two genera *Fenja* and *Aegir*, animals possessing the general form of sea anemones but distinguished from all Cœlenterates by the presence of a continuous and straight gut reaching from the mouth to the aboral pores which completely shuts off the cœlenteron or general body cavity from the stomodæum.

In more recent times the work has been by no means neglected. With the advantage of employing many modern improvements in the dredges and trawls in use, the American steamer, the 'Albatross,' has been engaged in a careful investigation of the deep-sea fauna of the eastern slopes of the Pacific Ocean, while at the same time Her Majesty's surveying vessel, the 'Investigator,' has been obtaining some interesting and valuable results from a survey of the deep waters of the Indian Ocean. But our knowledge of this vast and wonderful region is still in its infancy. We have gathered, as it were, only a few grains from a great unknown desert. It is true that we may not for many years, if ever, obtain any results that will cause the same deep interest and excitement to the scientific public as those obtained by the first great national expeditions, but there are still many important scientific problems that may be and will be solved by steady perseverance in this field of work, and if we can only obtain the same generous support from public institutions and from those in charge of national funds that we have received in the past two decades, many more important facts will doubtless be brought to light.

CHAPTER II

THE PHYSICAL CONDITIONS OF THE ABYSS

It is not surprising that the naturalists of the early part of the present century could not believe in the existence of a fauna at the bottom of the deep seas.

The extraordinary conditions of such a region—the enormous pressure, the absolute darkness, the probable absence of any vegetable life from want of direct sunlight—might very well have been considered sufficient to form an impassable barrier to the animals migrating from the shallow waters and to prevent the development of a fauna peculiarly its own.

The fragmentary accounts of animals brought up by sounding lines from great depths might, it is true, have thrown doubts on the current views; but they were not of sufficient importance in themselves, nor were the observations made with such regard to the possibility of error, as to withstand the critical remarks that were made to explain them away.

The absence of any evidence obtained by accurate systematic research, together with the consideration of the physical character of the ocean bed, were quite sufficient to lead scientific men of that period to doubt the existence of any animal life in water deeper than a few hundred fathoms.

We now know, however, that there is a very considerable fauna at enormous depths in all the great oceans, and we have acquired, moreover, considerable information concerning some of those peculiar physical conditions of the abyss that fifty years ago were merely matters of speculation among scientific men.

The relation between animals and their environment is now a question of such great interest and importance that it is necessary in any description of the fauna of a particular region to consider its physical conditions and the influence that it may be supposed to have had in producing the characteristics of the fauna.

The peculiar physical conditions of the deep seas may be briefly stated to be these: It is absolutely dark so far as actual sunlight is concerned, the temperature is only a few degrees above freezing point, the pressure is enormous, there is little or no

THE PHYSICAL CONDITIONS OF THE ABYSS 19

movement of the water, the bottom is composed of a uniform fine soft mud, and there is no plant life.

All of these physical conditions we can appreciate except the enormous pressure. Absolute darkness we know, the temperature of the deep seas is not an extraordinary one, the absence of movement in the water and the fine soft mud are conditions that we can readily appreciate; but the pressure is far greater than anything we can realise.

At a depth of 2,500 fathoms the pressure is, roughly speaking, two and a half tons per square inch—that is to say, several times greater than the pressure exerted by the steam upon the piston of our most powerful engines. Or, to put the matter in other words, the pressure per square inch upon the body of every animal that lives at the bottom of the Atlantic Ocean is about twenty-five times greater than the pressure that will drive a railway train.

A most beautiful experiment to illustrate the enormous force of this pressure was made during the voyage of H.M.S. 'Challenger.' I give the description of it in the words of the late Professor Moseley.

'Mr. Buchanan hermetically sealed up at both ends a thick glass tube full of air, several inches in length. He wrapped this sealed tube in flannel, and placed it, so wrapped up, in a wide copper tube, which was one of those used to protect the deep-sea thermometers when sent down with the sounding apparatus.

'This copper tube was closed by a lid fitting loosely, and with holes in it, and the copper bottom of the tube similarly had holes bored through it. The water thus had free access to the interior of the tube when it was lowered into the sea, and the tube was necessarily constructed with that object in view, in order that in its ordinary use the water should freely reach the contained thermometer.

'The copper case containing the sealed glass tube was sent down to a depth of 2,000 fathoms and drawn up again. It was then found that the copper wall of the case was bulged and bent inwards opposite the place where the glass tube lay, just as if it had been crumpled inward by being violently squeezed.

'The glass tube itself, within its flannel wrapper, was found when withdrawn, reduced to a fine powder, like snow almost. What had happened was that the sealed glass tube, when sinking to gradually increasing depths, had held out long against the pressure, but this at last had become too great for the glass to sustain, and the tube had suddenly given way and been crushed by the violence of the action to a fine powder. So violent and rapid had been the collapse that the water had not had time to rush in by means of the holes at both ends of the copper cylinder and thus fill the empty space left behind by the collapse of the glass tube, but had instead crushed in the copper wall and brought equilibrium in that manner. The process is exactly the reverse of an explosion, and is termed by Sir Wyville Thomson an "implosion."'

It is but reasonable to suppose that the ability to sustain this enormous pressure can only be acquired by animals after generations of gradual migrations from shallow waters. Those forms that are brought up by the dredge from the depths of the ocean are usually killed and distorted by the enormous and rapid diminution of pressure in their journey to the surface, and it is extremely probable that shallow-water forms would be similarly killed and crushed out of shape were they suddenly plunged into very deep water. The fish that live at these enormous depths are in consequence of the enormous pressure liable to a curious form of accident. If, in chasing their prey or for any other reason, they rise to a considerable distance above the floor of the ocean, the gases of their swimming bladder become considerably expanded and their specific gravity very greatly reduced. Up to a certain limit the muscles of their bodies can counteract the tendency to float upwards and enable the fish to regain its proper sphere of life at the bottom; but beyond that limit the muscles are not strong enough to drive the body downwards, and the fish, becoming more and more distended as it goes, is gradually killed on its long and involuntary journey to the surface of the sea. The deep-sea fish, then, are exposed to a danger that no other animals in this world are subject to, namely that of tumbling upwards.

That such accidents do occasionally occur is evidenced by the fact that some fish, which are now known to be true deep-sea forms, were discovered dead and floating on the surface of the ocean long before our modern investigations were commenced.

Until quite recently, every one agreed that no rays of sunlight could possibly penetrate the sea to a greater depth than a few hundred fathoms.

Moseley says that 'probably all is dark below 200 fathoms excepting in so far as light is given out by phosphorescent animals,' and Wyville Thomson speaks of the 'utter darkness of the deep-sea bottom.'

Within the last few years a few authors have maintained that it is quite possible that a few rays of sunlight do penetrate even to the greatest depths of the ocean—a view mainly based on the fact that so many deep-sea animals possess extremely perfect and complicated eyes and very brilliant colours. Verrill says: 'It seems to me probable that more or less sunlight does actually penetrate to the greatest depths of the ocean, in the form of a soft sea-green light, perhaps at 2,000 or 3,000 fathoms equal in intensity to our partially moonlight nights and possibly at the greatest depths equal only to starlight. It must be remembered that in the deep sea far away from land the water is far more transparent than near the coast.' Packard is of a similar opinion.

There seem to me to be very slight grounds for this view. The fact that, comparatively speaking, shallow-water fish avoid nets that are rendered phosphorescent by entangled jelly-fish does not justify us in assuming that deep-sea fish avoid regions where there are phosphorescent Gorgonians or Pennatulids. It is not by any means certain that fish avoid sunken nets on account of their phosphorescence. Most fish possess, as is well known, a very acute sense of smell, and it is very probable that they avoid such nets on account of the putrid odours of the dead animals that remain attached to them.

Nor is there much strength in the further argument that it can hardly be possible that there can be an amount of phosphorescent light regularly evolved by the few deep-sea animals, having this power, sufficient to cause any general illumination, or powerful enough to have influenced, over the whole ocean, the evolution of complex eyes, brilliant and complex protective colours, and complex commensal adaptations.

We have no sound information to go upon to be able to judge of the amount of light given off by phosphorescent animals at the bottom of the deep sea. The faint light they show on deck after their long journey from the depths in which they live to the surface may be extremely small compared with the light they give in their natural home under a pressure of $2\frac{1}{2}$ tons to the square inch. The complex eyes that many deep-sea animals exhibit were almost certainly not evolved as such, but are simple modifications of eyes possessed by a shallow-water ancestry.

THE PHYSICAL CONDITIONS OF THE ABYSS

The more recent experiments that have been made, tend to show that no sunlight whatever penetrates to a greater depth, to take an extreme limit, than 500 fathoms.

Fol and Sarasin, experimenting with very sensitive bromo-gelatine plates, found that there was no reaction after ten minutes' exposure at a depth of 400 metres on a sunny day in March.

But although it is highly probable that not a glimmer of sunlight ever penetrates to the depths of the ocean, there is in some places, undoubtedly, a very considerable illumination due to the phosphorescence of the inhabitants of the deep waters.

All the Alcyonarians are, according to Moseley, brilliantly phosphorescent when brought to the surface. Many deep-sea fish possess phosphorescent organs, and it is quite possible that many of the deep-sea Protozoa, Tunicates, Jelly-fish, and Crustacea are in their native haunts capable of giving out a very considerable amount of phosphorescent light.

If we may be allowed to compare the light of abysmal animals with that of surface forms, we can readily imagine that some regions of the sea may be as brightly illuminated as a European street is at night—an illumination with many very bright centres and many dark shadows, but quite sufficient for a vertebrate eye to distinguish readily and at a considerable distance both form and colour.

To give an example of the extent to which the illumination due to phosphorescent organisms may reach, I may quote a passage from the writings of the late Sir Wyville Thomson.

'After leaving the Cape Verde Islands the sea was a perfect blaze of phosphorescence. There was no moon, and although the night was perfectly clear and the stars shone brightly, the lustre of the heavens was fairly eclipsed by that of the sea. It was easy to read the smallest print, sitting at the after-port in my cabin, and the bows shed on either side rapidly widening wedges of radiance so vivid as to throw the sails and rigging into distinct lights and shadows.'

A very similar sight may frequently be seen in the Banda seas, where on calm nights the whole surface of the ocean seems to be a sheet of milky fire. The light is not only to be seen where the crests of waves are breaking, or the surface disturbed by the bows of the boat, but the phosphorescence extends as far as the eye can reach in all directions. It is impossible, of course, to say with any degree of certainty whether phosphorescence such as this exists at the bottom of the deep sea, but it is quite probable that it does in some places, and hence the well-developed eyes and brilliant colours of some of the deep-sea animals.

On the other hand the entire absence or rudimentary condition of the eyes of a very considerable proportion of deep-sea animals seems to prove that the phosphorescent illumination is not universally distributed, and that there must be some regions in which the darkness is so absolute that it can only be compared with the darkness of the great caves.

It is difficult to believe that the eyes of such animals as crabs and prawns for example would undergo degeneration if there were a glimmer of light in their habitat, a light even so faint as that of a star-light night in shallow water. With the faintest light the eyes would be of use to them in seeking their prey, avoiding their enemies, and finding their mates, and any diminution in the keenness of this sense would probably be of considerable disadvantage to them and tend to their ultimate extinction.

It might be argued that the animals of the abysses of the ocean probably feed chiefly upon the carcases of pelagic animals that have fallen from the upper regions of the sea, and that the sense of smell is probably the most important for them in searching for their food. That is quite probable; but many shallow-water animals invariably seek their food by their sense of smell without showing any traces of a weakness in their sense of sight. It may be taken as an axiom of biology that unless a particular sense is absolutely useless to an animal or a positive disadvantage to it, that sense will be retained.

It may be stated then with some confidence that in the abysmal depths of the ocean there is no trace of sunlight. It is highly improbable, on the face of it, that any ray of light could penetrate through a stratum of water four miles in thickness, even if the water were perfectly pure and clear, but when we remember that the upper regions, at least, are crowded with pelagic organisms provided with skeletons of lime and silica, we may justly consider that it is impossible.

The temperature of the water in the abyss is by no means constant for a constant depth nor does it vary with the latitude. It is true that, as a rule, the water is colder at greater depths than in shallower ones, and that the deeper the thermometer is lowered into the sea, the lower the mercury sinks. This is consistent with physical laws. If there is any difference at

THE PHYSICAL CONDITIONS OF THE ABYSS 29

all in the temperature of a column of water that has had time to settle, the thermometer will always reach its highest point at the top of the column and its lowest at the bottom, for the colder particles being of greater specific gravity than the warmer ones will sink, and the warmer ones will rise.

The truth of this will be clear if we imagine a locality at the bottom of a deep ocean with a source of great heat such as an active volcano.

Such a source of heat would, it is true, raise the temperature of the water in its immediate vicinity, but the particles of water thus heated would immediately commence to rise through the superjacent layers of colder water, and colder particles would fall to take their places. Thus the effect of an active volcano at the bottom of the deep sea would not be apparent at any very great distance in the same plane. In fact, unless the bottom of the ocean was closely studded with volcanoes we should expect to find, as indeed we do find, that the temperature of the sea rises as the water shallows.

If then we were to consider a great ocean as simply a huge basin of water, we should expect to find the water at the surface warmer than the water at the bottom. The temperature of the surface would vary constantly with the temperature of the air above it. That is to say, it would be warmer at the equator than in the temperate regions. The temperature at the bottom would be the same as the lowest temperature of the basin, that is, of the earth that supports it.

The great oceans however cannot be regarded as simple basins of water such as this. The temperature of the surface water varies only approximately with the latitude. It is generally speaking hottest at the equator and coldest at the poles, but surface currents in the intermediate regions produce many irregularities in the surface temperature.

Again, although we have no means of knowing what the temperature of the earth is at 1,000 fathoms below the surface of the ocean, it is very probable that in the great oceans the temperature of the deepest stratum of water is considerably lower than the true earth temperature. This is due to currents of cold water constantly flowing from the poles towards the equator. If these polar currents were at any time to cease, the temperature of the lowest strata of water would rise.

Although the polar currents cannot be actually demonstrated nor their exact rapidity be accurately determined, the deduction from the known facts of physical geography that they do actually exist is perfectly sound and beyond dispute. A few considerations will, I think, make this clear.

If the ocean were a simple basin somewhat deeper at the equator than at the poles, the cold water at the poles would gradually sink down the slopes of the basin towards the latitude of the equator, and the bottom temperature of the water would be constant all the world over.

A few hills here and there would not affect the general statement that for a constant depth the temperature of the lowest stratum of water would be constant.

But in some places ridges occur stretching across the oceans from continent to continent, and these ridges shut off the cold water at the bottom of the sea on the polar side from reaching the bottom of the sea on the equator side.

If A (fig. 1) represents a ridge stretching from continent to continent across an ocean, and the arrow represents the direction of the current, then the water that flows across the ridge from the polar side to the equator side will be drawn from the layers of water lying above the level of the ridge, and consequently none of the coldest water will ever get across it, and from the level of the ridge to the bottom of the sea on the equatorial side the water will have the same temperature as the water at the level of the ridge on the polar side.

It follows from this that in places where there are deep holes in the bed of the ocean surrounded on all sides by considerable elevations, the temperature of

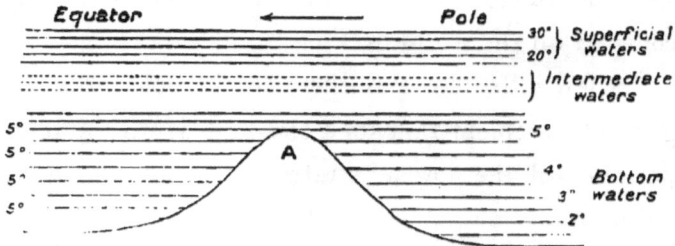

FIG. 1.—Diagram illustrating the passage of an ocean current across a barrier (A).

the water at the bottom will be the same as the temperature of the water on the summit of the lowest ridges that surrounds them.

This explains why it is that we find that the bottom temperature for a given depth is frequently less in one place than it is in another, even in places of the same parallel of latitude. One or two examples may be taken to illustrate these points. The temperature off Rio Janeiro in lat. 20° S. was found by the

'Challenger' to be 0·6° C. at a depth of 2,150 fathoms. In a similar latitude north of the equator at a depth of 2,900 fathoms the temperature was found to be 2·2° C., and at a point near Porto Rico there is a deep hole of 4,561 fathoms with a bottom temperature of 2·2° C.

Again it has been shown by the American expedition that the temperature of the water at the deepest point in the Gulf of Mexico, 2,119 fathoms, is the same as that of the bottom of the Straits of Yucatan, 1,127 fathoms, namely 4·1° C. And, passing to another part of the world altogether, we find in the small but deep sea that lies between the Philippines and Borneo that, at a depth of 2,550 fathoms, the temperature is 10·2° C.

These facts then show that, although at the bottom of the deep seas the water is always very cold, the degree of coldness is by no means constant in the same latitude for the same depth.

We must now return to the polar currents. We have assumed above that these currents do exist, and it is probable that by this time the reader must have seen why they are assumed to exist.

The water at the bottom of the ocean is exceedingly cold. Where does this coldness come from? It is obvious that in temperate and tropical climes it does not come from the surface. Nor is it at all probable that it comes from the earth upon which the water rests; for, if it were so, the temperature for water of a given depth would always be the same. We should not find the bottom temperature of 0·4° C. at 2,900 fathoms off Rio de la Plata and a temperature of 2·2° F. in 4,561 fathoms off Porto Rico.

In fact the only hypothesis that can with any show of reason be put forward to account for the temperature of the bottom of the ocean is that which derives its coldness from the Polar ice.

We have at present very little evidence to enable us to judge of the force and direction of the polar currents in the two hemispheres, but the researches of the 'Challenger' prove almost conclusively that in the Atlantic Ocean there is a very strong predominance of the Antarctic polar current. In fact it seems very probable that the Arctic polar current, if it exist at all, is very small and confined to the eastern and western shores of the North Atlantic.

It is very probable, however, that these currents at the bottom of the ocean are extremely slow, and, as the water is never affected by tides or storms, the general character of the deep sea is probably one of calm repose. This is a matter of no little importance; for, in the consideration of the characters presented by the fauna of any particular region, it is always necessary to take into account the physical difficulties the animals have to contend against and the modifications of structure they present to combat these difficulties.

Thus in a region such as that presented by the deep sea, where there are no rapid tides, we should not expect to find such a powerful set of body muscles in the free-swimming forms nor such a firm vertebral column as in the animals that live in more lively water.

Perhaps it is of the nature of an assumption to say that there are no rapid currents and tides in the abysmal depths of the ocean, for we have no means of demonstrating or even of calculating the rate of flow of these waters. But it is a reasonable hypothesis and one that we may well use until the contrary is proved.

A fact of some importance that supports this hypothesis, as regards some parts of the ocean at least, is presented by the sea-anemones.

Many of the shallow-water Actinians are known to possess minute slits in the tentacles and disc, affording a free communication between the general body cavity or cœlenteron and the exterior.

In many deep-sea forms the tentacles are considerably shorter and the apertures larger than they are in shallow-water forms. It is difficult to believe that such forms, perforated by, comparatively speaking, large holes, could manage to live in rapidly flowing water, for if they did so they would soon be smothered

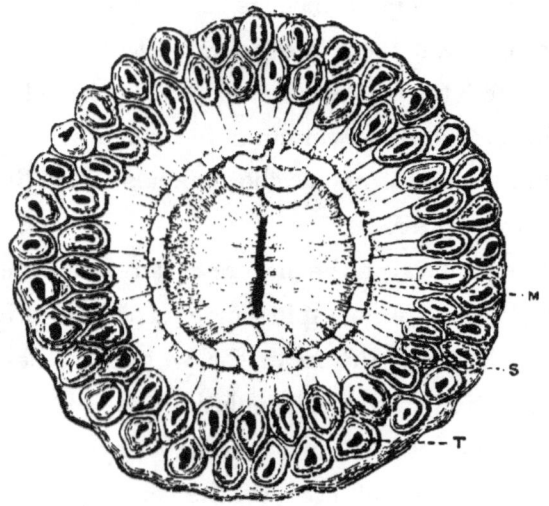

FIG. 2.—*Sicyonis crassa.* M, mouth; S, ciliated groove; T, tentacles. Each tentacle is perforated by a single large aperture. (After Hertwig.)

by the fine mud that composes the floor of all the deep seas. In fact anemones of the type presented by such forms as *Sicyonis crassa* are only fitted for existence in sluggish or still water.

Another character that must be taken into con-

THE PHYSICAL CONDITIONS OF THE ABYSS 37

sideration is that presented by the floor of the great oceans. The floor of the ocean, if it were laid bare, would probably present a vast undulating plain of fine mud. Not a rock, not even a stone would be visible for miles.

The mud varies in different parts of the globe according to the depth, the proximity to land, the presence of neighbouring volcanoes or the mouths of great rivers.

The Globigerina ooze is perhaps the best known of all the different deep-sea deposits. It was discovered and first described by the officers of the American Coast Survey in 1853. It is found in great abundance in the Atlantic Ocean in regions shallower than 2,200 fathoms. Deeper than this, it gradually merges into the 'Red mud.' It is mainly composed of the shells of Foraminifera, and of these the different species of Globigerina are the most abundant. It is probably formed partly by the shells of the dead Foraminifera that actually live on the bottom of the ocean and partly by the shells of those that live near the surface or in intermediate depths and fall to the bottom when their lives are done.

So abundant are the shells of these Protozoa that nearly 95 per cent. of the Globigerina ooze is composed of carbonate of lime. The remaining five per cent. is composed of sulphate and phosphate of lime, carbonate of ammonia, the oxides of iron and manganese, and argillaceous matters. The oxides of iron and manganese are probably of meteoric origin; the argillaceous matter may be due to the trituration

FIG. 3.—Globigerina ooze. (After Agassiz.)

of lumps of pumice stone and to the deposits caused by dust storms.

Globigerina ooze may be found on the floor of the ocean at depths ranging from 500 to 2,800 fathoms of water in equatorial and temperate latitudes. The reason that it is not found in Arctic seas may be that the cold surface waters of these regions do not bear such an abundant fauna of Foraminifera. This is supported by the fact that it extends ten degrees further north than south in the Atlantic, the warm water of the Gulf Stream bearing a richer fauna than the waters of a corresponding degree of latitude in the Southern Sea.

The Pteropod ooze has only twenty-five per cent. of carbonate of lime. It contains numerous shells of various Pteropods, Heteropods, and Foraminifera, but nearly fifty per cent. of its substance is composed of the siliceous skeletons of Radiolaria and the frustules of diatoms.

According to Murray it is found in tropical and subtropical seas at depths of less than 1,500 fathoms.

The Radiolarian ooze is found only in the deepest waters of the Central and Western Pacific Ocean. In some of the typical examples, not a trace of carbonate of lime was to be found, but in somewhat shallower waters a few small fragments occurred.

A Diatom ooze, mainly composed of the skeletons of diatoms, has also been found in deep water near the Antarctic Circle, but it has not apparently a very wide range.

Of all the deep-sea deposits, however, the so-called 'Red mud' has by far the widest distribution.

It is supposed to extend over one-third of the earth's surface. It is essentially a deep-sea deposit, and one that is found in its typical condition at some considerable distance from continental land. Like the Globigerina ooze it is never found in enclosed seas. To the touch it is plastic and greasy when fresh, but it soon hardens into solid masses. When examined with the microscope it is seen to be composed of extremely minute fragments rarely exceeding 0·05 mm. in diameter. It contains a large amount of free silica that is probably formed by the destruction of numerous siliceous skeletons, and a small proportion of silicate of alumina. It usually contains the remains of diatoms, radiolaria, and sponge spicules, and occasionally lumps of pumice stone, meteoric nodules, and, in colder regions, stones and other materials dropped by passing icebergs.

In the great oceans, then, we find in the deepest places red mud, or, where there is an abundant radiolarian surface fauna, Radiolarian ooze; in water that is not deeper than about 2,000 fathoms, we find the Globigerina ooze; in shallower waters and in some localities only Pteropod ooze.

It must not be supposed that sharp limits can anywhere be drawn between these different kinds of deposits, for they pass gradually into one another and present many intermediate forms.

It is probable that the sea-water by virtue of the free carbonic acid it contains in solution is able to exert a solvent action upon the calcium carbonate shells of animals as they sink to the bottom, and during the long and very slow journey from the surface to the bottom of the deepest seas these shells are completely dissolved.

The first to be dissolved would be the thin delicate shells of the Pteropods and Heteropods, for besides the fact that they present a wider surface to the solvent action of the water they are probably influenced more by tide and currents, sink more slowly and erratically, and thus have a longer journey to perform.

Then the smaller but more solid and compact shells of the Foraminifera are dissolved, and lastly, in the deepest water only the siliceous skeletons of the radiolaria and diatoms are able to reach their last resting place at the bottom of the ocean.

These four oozes then are characteristic of the floor of the deep oceans. In the proximity of land and in inland seas where deep water occurs, other muds are found differing from one another in accordance with the character of the coasts in their vicinity. It is not necessary to give a detailed account of them, but a few remarks on some of the more pronounced forms may not be without interest.

The blue mud contains eighty per cent. of a mixture of quartz, mica, felspar, hornblende, and other minerals, mixed with a considerable quantity of decomposing animal and vegetable substance, the calcareous remains of foraminifera, mollusca, worms, echinoderms, alcyonaria and corals, and the siliceous skeletons of radiolaria and diatoms.

The green mud is characterised by a large percentage of glauconite.

The red muds characteristic of the Brazilian coast contain a large amount of ochreous matter brought into the sea by the great rivers.

In the neighbourhood of active volcanoes there is a characteristic volcanic mud, and in the coral seas the deep-sea deposits contain a large percentage of the calcareous remains of dead corals.

One more character of the deep-sea region must be referred to before we pass on, and that is the absence of vegetable life. It has not been determined yet with any degree of accuracy where we are to place the limit of vegetable life, but it seems probable that below a hundred fathoms no organisms, excepting a few parasitic fungi, are to be found that can be included in the vegetable kingdom. While then the researches of recent times have proved beyond a doubt that there is no depth of the ocean that can be called azoic, they have but confirmed the perfectly just beliefs of the older naturalists that there is a limit where vegetable life becomes extinct. It is not difficult to see the reason for this. All plants, except a few parasites and saprophytes, are dependent upon the influence of direct sunlight, and as it has been shown above that the sunlight cannot penetrate more than a few hundred fathoms of sea-water, it is impossible for plants to live below that depth.

The absence of vegetable life is an important point in the consideration of the abysmal fauna, for it is in consequence necessary to bear in mind that the food of deep-sea animals must be derived from the surface. It is possible that deep-sea fish, in some cases, feed upon one another and upon deep-sea crustacea, that deep-sea crustacea feed upon deep-sea worms, that deep-sea echinoderms feed upon deep-sea foraminifera, and so on through all the different combinations; but the fauna would soon become exhausted if it had no other source of food supply. This other source of food supply is derived from the bodies of pelagic organisms that fall from the upper waters of the ocean, and is composed of protozoa, floating tunicates, crustacea, fish, and other animals, together with diatoms and fragments of sea-weed.

CHAPTER III

THE RELATIONS OF THE ABYSMAL ZONE AND THE ORIGIN OF ITS FAUNA

In the study of the geographical distribution of terrestrial animals one of the great difficulties met with is the impossibility of defining exactly the limits of the regions into which we divide the surface of the earth. In a general way we recognise that there is an Australian region, an Ethiopian region, &c.; but, when we come to discuss the exact position of the frontier lines that separate these regions from their neighbours, we find all kinds of difficulties to overcome and inconsistencies to meet.

For the sake of convenience it is useful to adopt certain arbitrary limits for these regions, notwithstanding these difficulties and inconsistencies, but we must recognise the fact that nature recognises no such limits, that every region overlaps its neighbours to a greater or less extent, and that there are many debateable grounds in the world where the fauna characteristic of one region is mixed with that characteristic of another.

But this difficulty in defining the exact limits of the terrestrial faunistic regions is even more pronounced in the case of the regions and zones of the marine fauna.

On the dry land we find mountain ranges, forests, deserts, and other barriers, that to a very considerable extent prevent the mixing of one fauna with another, but in the sea there are no barriers of anything like the same importance, but one fauna gradually merges into the neighbouring fauna according to the temperature, the pressure, the amount of light, the salinity of the water or the food supply. This then is one of the difficulties met with in the study of the geographical distribution of the marine fauna.

But there is another that leads to almost greater complications. In considering terrestrial life it is customary to refer only to regions of geographical, or perhaps it would be more correct to call it—superficial distribution. It would be quite possible, however, to subdivide the geographical areas into zones of elevation above the sea-level, not very clearly marked off from one another, it is true, but nevertheless each showing a number of characteristic features. This idea is expressed, for example, when we speak of the Alpine fauna, the Himalayan fauna, or the fauna of the great Andes.

In the study of the marine fauna and flora we must notice, it is the depth of the water, or in other words the depression of the habitats below the sea-level, that forms the most important consideration. Geographical sub-regions may be recognised and defined with a certain amount of accuracy, especially in the case of the fauna of the shallow waters, but by far the most important changes in the general characters of the fauna are found when we pass from one ' zone ' of depression to another. Thus in describing any particular marine fauna we should mention first of all its zone or sub-zone of depression and then its geographical region and sub-region. For example, we may speak of the fauna of the pelagic zone of the British sub-region of the European region, or the fauna of the abysmal zone of the Northern sub-region of the Atlantic region.

We can recognise three primary zones of the marine fauna which we may call the 'Pelagic,' the ' Neritic,' and the 'Abysmal' zones.

The Pelagic zone includes the superficial waters

of all seas extending from the surface to a depth which cannot at present be very accurately determined, but is probably the same as the limit of the influence of direct sunlight.

The animals of this zone are frequently characterised by a general transparency of their tissues, a white or sea-water (i.e. blue or green) colour, an organisation capable of prolonged swimming or floating movement, and by giving birth to floating eggs which hatch out transparent larvæ or embryos.

The pelagic zone may be divided into several geographical regions and sub-regions, which it would be beyond the scope of this book to enumerate here, but there is one that calls for a few brief remarks. In many parts of the ocean there may be found vast areas of floating sea-weed, which carry with them a population of crustacea and other animals peculiarly their own. This 'sargasso' fauna presents so many characteristics and so many features different from that of the ordinary pelagic fauna, that the tracts of sea bearing this weed must be considered to rank as a special region of the pelagic zone, which may be called the Sargasso region.

The zone of shallow water for which we shall adopt Professor Haeckel's term—the Neritic zone—embraces all parts of the seas of less depth than 500 fathoms, including the inland seas, the shores of great continents and islands, and the shallow banks in the great oceans. It does not include the superficial waters—which belong to the pelagic zone—but extends only from the actual bottom to a distance of a few fathoms above it. The fauna of this zone is extremely varied, consisting of animals that swim, crawl, or are permanently fixed to the bottom, animals of almost every variety of colour and marking, and of every size and shape.

The exact limits of the Neritic sub-zones are not easy to define. The distinguished naturalist Forbes, to whom the abysmal zone was unknown, divided the seas from 0–50 fathoms in depth into three zones —the littoral zone lying between tide marks, the laminarian zone extending from 0–15 fathoms, and the coralline zone from 15–50 fathoms.

The first of these will stand as a sub-zone, the animals that are able to withstand exposure to the sun and air either in pools or upon the rocks and sand even for a few minutes frequently possessing features that distinguish them from those dwelling beyond low-water mark, just as those more active creatures that migrate backwards and forwards with the ebb and flow of every tide differ from the dwellers in the open sea. There is, it is true, at every low tide, a migration of part of the fauna of this sub-zone into the next, but still it is sufficiently well defined to be allowed to remain in our category.

The second sub-zone is not so easy to define. The terms 'laminarian' and 'coralline' used by Forbes are only applicable to certain geographical regions and must be abandoned for general use.

We can only recognise one sub-zone between the littoral sub-zone and the abysmal zone, for notwithstanding the important varieties it exhibits in the nature of the bottom, whether it be rocky, sandy, or weedy, the amount of light, the temperature of the water, and the rapidity of the currents, it is not possible at present to point to any general characters of the fauna of its different parts to justify us in subdividing it.

The name that may be given to this second sub-zone of the neritic zone is the Katantic—the sub-zone of the slopes.

The last well-marked zone is the abysmal, extending from the 500-fathom line to the greatest depths of the ocean, one of enormous superficial area, one that it is most difficult to investigate, and one about which we know but little.

In the present state of our knowledge we cannot divide it into any well-marked sub-zones nor even into geographical regions or sub-regions. It is not divided into sections by any important geographical barriers, and the general characters presented by its fauna are practically the same all the world over.

Professor A. Agassiz has pointed out in his 'Challenger' monograph that the deep-sea echinoids of the Atlantic Ocean differ from those living in corresponding depths in the Pacific Ocean, but it is doubtful whether any such well-marked differences can be observed in other groups of animals. If, in the course of time, increased knowledge of deep-sea animals emphasises the difference between the abysmal fauna of the Pacific and that of the Atlantic, then we can divide this zone into two geographical regions; but at present it seems more correct to consider the abysmal zone as one that is indivisible either bathymetrically or geographically.

Before passing on to the consideration of the general characters of the abysmal fauna, there are still one or two points that must be just briefly referred to.

It is the function of every true naturalist to consider animals from every possible point of view. Not only must he regard them as members of a certain species belonging to a genus, a family, an order, and so on, presenting certain peculiarities of structure and development; not only must he regard them as inhabitants of a certain locality or zone of depth, but he must also pay attention to their habits and mode of life.

Now amongst marine animals we can recognise three principal modes of life. Some animals simply float or drift about with the currents of the sea and are unable to determine for themselves, excepting, perhaps, within very small limits, the direction in which they travel. Such are the countless forms of protozoa, the jelly-fishes and medusæ, numerous pelagic worms and crustacea, the pyrosomas and salps, and many other forms well known to those who are in the habit of using the tow-net. This portion of the fauna has recently been called the Plankton.

Then there are the animals that are capable of very considerable swimming movements, animals that are able to stem the tide and migrate at will from one part of the sea to another, such as the cetacea, most fishes, and perhaps also many cephalopods. This portion of the fauna has been called the Nekton.

And lastly we have those animals that remain perfectly fixed to the bottom or are capable only of creeping or crawling over the rocks and sand, such as the sponges, hydroids, sedentary tunicates, gasteropods, most lamellibranchs, and many crustacea. This portion of the fauna has been called the Benthos.

Although it will not be necessary to use these terms very frequently in this little book, it may be advisable for the reader to bear in mind that in any exhaustive treatise on the marine fauna such terms would be employed, and that in the chapters dealing with the fauna of the abysmal zone we should find accounts of the 'bathybial plankton,' the 'bathybial nekton,' and the 'bathybial benthos.'

Lastly we must consider quite briefly the views that have been held concerning the origin of the abysmal fauna.

As soon as it became clear to naturalists that there is no part of the ocean, however deep it may be, that deserves the name 'azoic,' but that almost every part has a fauna of greater or less density, the problem of the origin of this fauna presented itself.

Whence came the curious creatures that live mostly in total darkness and can sustain without injury to their delicate and complicated organisation the enormous pressure of the great depths? Are they the remnants of the fauna of shallow prehistoric seas that have reached their present position by the gradual sinking of the ocean basins? Or, are we to look upon the abysmal region as the nursery of the marine fauna, the place whence the population of the shallow waters was derived? Neither of these answers is supported by the facts with which we are now well acquainted. The fauna of the abysmal region does not show a close resemblance to that of any of the past epochs as revealed to us by geology, nor are we justified in assuming without much stronger evidence than we now possess, that the oceans have undergone any such great depression as this first theory presupposes.

Nor can we consider for a moment that the abyss was the original source of the shallow-water fauna; for not only do we find but few types that can be considered to be, in any sense of the word, ancestral in character; but on the contrary most of the animals of the deep sea seem to be specially modified types of shallow-water forms. The most probable explanation of the origin of the deep-sea fauna is the one that was put forward by Moseley and has been since supported by almost every authority on the subject, namely, that the fauna of the deep sea has been derived from successive immigrations of the animals from the shallow water.

This view is supported by the fact that the deep-sea fauna is much richer in the neighbourhood of land than it is in regions more remote from it. Many examples could be given to illustrate this point. The extraordinary richness of the deep-sea fauna on the western slopes of the floor of the Atlantic has been frequently commented on by the naturalists connected with the expeditions of the American vessels, the 'Blake,' the 'Fish Hawk,' and the 'Albatross.' Moseley called attention several years ago to a few localities in the neighbourhood of the land especially rich in deep-sea forms in comparatively shallow waters, such as one near the island of Sombrero in the Danish West Indies, where within sight of the lighthouse a haul of the dredge in 450 fathoms brought up a rich fauna of blind crustacea, corals, echinoderms, sponges, &c. Another off Kermadec in 630 fathoms brought up numerous curious blind fishes, ascidians, cuttlefishes, crustacea, *Pentacrinus*, and large vitreous sponges, and there are similar localities lying between Aru and Ke and between the Nanusa archipelago and the Talaut islands. The deep water off the Norwegian, Scotch, Irish, and Portuguese coasts also seems to be particularly rich in various forms of animal life. The same is probably true of the deep sea of many other regions in the neighbourhood of land, and, although it cannot be taken to be a rule without exceptions—the abysmal fauna off the western coasts of the Panama region being, according to the recent researches of Alexander Agassiz in the 'Albatross,' particularly poor—yet we can assert as a statement of very general application that the further removed from continental land, the poorer is the abysmal fauna.

Another argument that has been brought forward by Moseley in support of his view is that there is a certain relationship between the deep-sea fauna of any particular region and the shallow-water fauna of the nearest coasts. This is a point that is not easy to illustrate by examples, but as Moseley's argument has not, so far as I am aware, been disputed by any of the naturalists who have followed him in this line of work, and the recent results of the 'Albatross' in comparing the deep-sea fauna of the eastern and western sides of the isthmus of Panama seem if anything to support it, we can take it as a point in favour of his view of the origin of the abysmal fauna.

It is impossible to say at present at what time in the world's history these migrations commenced, but, as Agassiz points out, none of the palæozoic forms are found in the deep sea, and this seems to indicate that the fauna did not commence its existence earlier than the cretaceous period.

It is quite possible, however, that part of the fauna of the deep sea has been derived directly from the pelagic zone. The occurrence of bathybial Radiolaria, Foraminifera and Siphonophora, and among fishes genera and species of the pelagic families Sternoptychidæ and Scopelidæ, suggest that this zone may have contributed very largely to the fauna of the abyss.

Much of course still remains to be done before we can consider any of these interesting problems connected with the deep-sea fauna to be definitely solved. All we can do at present is to speculate upon the direction in which the facts at our disposal seem to point, and by following up one clue after another hope that we may eventually arrive at the truth. The task may be a difficult one, but it will reward our efforts. If truth is hard to find when it lies at the bottom of a well, how much more inaccessible must it be when it lies hidden in the darkness of the sea's abyss!

CHAPTER IV

THE CHARACTERS OF THE DEEP-SEA FAUNA

THE general characters presented by animals living in deep water may be considered under several headings. The most important are those that are directly or indirectly related to the fact that the animals live either in total darkness or in the faint and probably intermittent light emitted by phosphorescent animals; namely, the colour of the skin and the peculiarities of the eyes.

The colours of the skin of the deep-sea animals vary to a very remarkable extent in the different groups. It cannot be said that there is any one colour at all predominant, and it is only in certain classes that black, white, or dull-coloured animals are more numerous than others. The colours are however usually very evenly distributed, and we find but few examples of animals with spots, stripes, or other pronounced markings.

The majority of the fish are dark brown or black, but many other colours are represented. Thus

Ipnops Murrayi, a typical deep-sea fish, is yellowish brown with colourless fins, and it exhibits a further character not uncommon in these abysmal forms, namely black buccal and branchial cavities. *Typhlonus nasus*, again, is said to be of a light brownish colour, with black fins. Many other examples could be given to show the prevalence in these regions of these black, dull, and pale uniform colours. But there are many exceptional cases. *Neoscopelus macrolepidotus*, for example—a form that according to Günther undoubtedly belongs to the bathybial region—is distinguished by its brilliant colours. It is bright red mixed with azure blue, the whole relieved by silver spots with circles of black on the abdomen.

Prorogadus nudus is of a pale rose colour, with the under and lateral sides of the head bluish black.

Rhodichthys regina, found in 1,280 fathoms of water, is uniformly bright red in colour.

A. Agassiz says in his reports on the dredging operations on the west coast of America: 'The coloration of the deep-sea fishes is comparatively monotonous. The tints are all a light violet base, tending more or less to brownish or brownish yellow, or even to a greenish tint, especially among the Macruridæ. Some of the Liparidæ were of a dark violet, and one species was characterised by a brilliant blue band. The Ophidiidæ, *Nemichthys*, and the like, are usually of an ashy violet tint, while in *Ipnops* and *Bathypterois* the tints were of a decidedly yellowish brown.'

That the deep-sea fish are usually devoid of any pronounced spots, stripes, and other markings, is now well recognised. It may not be altogether out of place, however, to refer briefly to a few exceptions.

The black circles on the abdomen of *Neoscopelus macrolepidotus* have already been referred to.

Halosaurus johnsonianus has a black spot on the tail.

Aulostoma longipes has three pairs of large black spots on the ventral side, but the specimen taken in 1,163 metres of water by the 'Talisman' was probably a young one.

It is very probable that in all the exceptional cases, when fish taken in deep-sea water have exhibited such spots and markings, they are examples either of fish that have quite recently adopted an abysmal habitat or of young specimens exhibiting ancestral inherited characters.

In referring to a specimen of *Raja circularis*, taken by the 'Triton' in 516 fathoms, Günther says: ' It is notable that the spot on each side of the back which in littoral specimens is variegated with yellow is much smaller in the deep-sea specimen and uniformly black without yellow.'

It seems to be then a very general rule among fishes that as they migrate into deeper water the spots and stripes, so conspicuous among many forms living on the surface and in shallow water, disappear, and the coloration of the body becomes more evenly distributed and uniform.

Among the Mollusca, the deep-sea Cephalopods seem to be usually violet, but an *Opisthoteuthis Agassizii* caught by the ' Blake ' is stated to be of a dark chocolate colour, a *Nectoteuthis Pourtalesii* reddish-brown, and a *Mastigoteuthis* orange brown, while of the specimens brought home by the 'Challenger,' *Cirroteuthis magna* was said to be 'rose' when captured, and the spirit specimens of *Cirroteuthis pacifica* and *Bathyteuthis abyssicola* were purplish madder and purplish brown respectively.

The shells of the Gasteropods and Lamellibranchs living in the abyss are frequently so thin as to be almost transparent, and are, with very few exceptions, white or pale straw coloured. The colour of the only specimen of nudibranchiate Mollusca that has been found in the abysmal zone, namely, *Bathydoris abyssorum*, is described by Mr. Murray as follows: 'The body of the living animal was gelatinous and transparent, the tentacles brown, the gills and protruding external generative organ orange, the foot dark purple.'

Among the Crustacea various shades of red are the prevailing colours. 'The deep-sea types, like *Gnathophausia*, *Notostomus*, and *Glyphocrangon*,' says Agassiz, 'are of a brilliant scarlet; in some types, as in the Munidæ and Willemoesiæ, the coloration tends to pinkish or yellowish pink, while in *Nephrops* and *Heterocarpus* the scarlet passes more into greenish tints and patches.'[1] But perhaps the most remarkable point in the colour of the crustacea is that which immediately follows the paragraph I have just quoted. 'The large eggs of some of the deep-sea genera are of a brilliant light blue, and in one genus of Macrura we found a dark metallic blue patch on the dorsal part of the carapace in marked contrast to the brilliant crimson of the rest of the body.'

[1] In the recent researches of the 'Investigator' a few crustacea of rather exceptional colour were found. Whilst the great majority of them are described to be pink or red in colour when alive, *Gnathophausia bengalensis* is deep purple lake, *Haliporus neptunus* lurid orange, and *Aristaeus coruscans* bright orange.

The occurrence of this blue colour in Crustaceans of the deep sea is very remarkable, for blue is a colour, as Moseley pointed out many years ago, that is rarely met with in the fauna of the abyss, and it is certainly very exceptional in the crustacea of that zone.

Among the deep-sea Echinoderma we find a wonderful variety of coloration. Moseley says that many deep-sea Holothurians, for example, are deep purple, and Agassiz reports that in one species the colour was of a delicate green tinge. 'We obtained,' he adds, 'a white *Cucumaria* and some species of *Benthodytes* of the same colour,' while others vary from transparent milky white to yellow and light yellowish brown and even pinkish colours. The Crinoids are described by the authorities to be white, purple, yellow and brownish-chestnut, and of the other groups of the echinoderms we read that the star-fishes are, as a rule, of duller colours than the crustacea, but all more or less pink or red. 'The Hymenasteridæ, on the contrary, vary from light bluish violet to deep reddish chestnut colours.' The brittle stars are red and orange or dullish grey, while the urchins may be deep violet, claret coloured, brownish, or of a delicate pink.

It is impossible to account for this extraordinary

variety of colour in the deep-sea echinoderms. It is hardly probable that it can be protective or warning in function, and it is difficult to suppose that it is due to any peculiar excretory process. Whether it is due in any way to the influence of the environment, or, like the colour of autumn leaves, to the chemical degeneration of colours that in the shallow-water ancestry were functional, are problems that must be left for the future to decide.

The colour of the deep-sea Cœlenterates has unfortunately not been recorded in all cases, but still the few observations that we have, show that in this group, as in the last, almost every tint and shade are represented.

The colouring of the deep-sea jelly-fishes is said to be usually deep violet or yellowish red. However 'a species of *Stomobrachium*,' says Agassiz, 'is remarkable for its light carmine colour, a tint hitherto not observed among Acalephs.'

Moseley records most minutely the colour of some of the deep-sea anemones and corals, and calls attention to the very general presence of madder brown in the soft parts. Agassiz says: 'Among deep-sea Actiniæ, a species of a new *Cereanthus* was of a dark brick-red, while other actinians allied to

Bunodes were of a deep violet. Actinauge-like forms with tentacles of a pinkish-violet tinge frequently have the column of a yellow shade. The Zoanthidæ were greyish-green.' And again, in his narrative of the voyage of the 'Blake,' he records that 'some of the deep-sea corals are scarlet, deep flesh-coloured, pinkish orange, and of other colours,' and in referring to the Gorgonian *Iridogorgia* he says: 'The species are remarkable for their elegance of form and for the brilliant lustre and iridescent colours of the axis, in some of a bright emerald green, in others like burnished gold or mother-of-pearl.'

The fauna of the deep sea then, taken as a whole, is not characterised by the predominance of any one colour. The shades of red occur rather more frequently than they do in the fauna of any other zone or region, but whether this is in any way connected with the fact that red is the complementary colour to that of the phosphorescent light, in which many of these animals live, it is at present difficult to say; it is possible that, when we have further information concerning the colours of the animals living in the deeper parts of the Neritic zone, another explanation may be forthcoming.

Moseley points out that there are no blue animals

THE CHARACTERS OF THE DEEP-SEA FAUNA 67

known to live in deep water, and it might be added that green is extremely rare as a colouring matter in abysmal animals, although the phosphorescent light given out by some of the echinoderms is green.

Blue, as a colouring matter of marine animals, living on the surface or in shallow water, is not uncommonly met with, distributed in the form of bands or stripes, but green is extremely common in fishes, crustacea and coelenterates, and it is a point of very considerable importance that in this respect there is a very great difference between the deep-sea and the shallow-sea faunas.

If a considerable collection of living abysmal forms could be placed upon one table and a similar collection of shallow-water forms upon another, I believe that the first general impression upon the mind of one who saw them both for the first time would be the presence of green colours in the last-named collection, and the absence of it in the other.

The eyes of the animals that live in deep-sea water undergo curious modifications. If the fauna of the abysmal region were confined to conditions of absolute darkness, we should expect to find either a total absence of eyes or mere rudiments of them only in those forms that have recently migrated from the shallow water. This is the case with the fauna of the great caves. There is probably total darkness in these underground lakes and streams, and there is only the remotest possibility of the animals living in them ever seeing, even temporarily, a ray of sunlight or even a glimmer of phosphorescence during the whole of their life-time. We find then that the cave fauna is totally blind.

The conditions in the deep sea are not quite the same. In some regions there is probably a very considerable illumination by phosphorescent light, and it is quite possible that many of the characteristic deep-sea forms may occasionally wander into shallower regions where faint rays of sunlight penetrate, or even that the young stages of some species may be passed at or near the surface of the sea. Taking these points into consideration, then, it is not surprising to find that, in the deep seas, there are very few animals, belonging to families usually provided with eyes, that are quite blind.

In the majority of cases we find that the eyes are either very large or very small. Only in a small minority of cases do we find that the eyes are recorded to be moderate in size. The relation between the large-eyed forms and the small-eyed forms is not the same in all the regions of deep seas. In depths of 300 to 600 fathoms the majority are large-eyed forms. This is as we should expect, for it is more than probable that many of these forms occasionally wander into shallower waters where there is a certain amount of sunlight.

In depths of over 1,000 fathoms, the small-eyed and blind forms are in a majority, although many large-eyed forms are to be found.

Among fishes, for example, we find the species of *Haloporphyrus* found in depths of 300–600 fathoms with large eyes; and so with *Dicrolene*. *Cyttus abbreviatus*, and many other forms that are known to live in water of less depth than 700 fathoms; while on the other hand in *Melanocetus Murrayi*, *Ipnops Murrayi*, many deep-sea eels and other fish that are truly abysmal and live chiefly in depths of over 1,000 fathoms, the eyes are either very small or absent.

Some interesting examples may be found in the species of widely distributed genera to illustrate these points. Thus in *Neobythites grandis*, from 1,875 fathoms, the eye is small, only one-eleventh the length of the head, but in *Neobythites macrops*, *N. ocellatus*, and *N. gillii* from shallower water it is much larger.

N. grandis .	1,785 fms.	Eye $\frac{1}{11}$th length of the head	
,, macrops .	375 ,,	,, $\frac{2}{9}$	,, ,,
,, ocellatus	350 ,,	,, $\frac{1}{4}$	,, ,,
,, gillii	111 ,,	,, $\frac{1}{33}$	,, ,,

Similarly in the species of the widely distributed deep-sea genus *Macrurus*: the species *M. parallelus, japonicus, M. fasciatus,* &c., usually living in water less than 1,000 fathoms deep, have large and in some cases very large (*M. fasciatus*) eyes, but *Macrurus filicauda, M. fernandezianus, M. liocephalus, M. Murrayi, M. armatus* have small eyes.

Some deep-sea fish have their eyes reduced to a mere rudiment; such as *Ceratias uranoscopus, C. carunculatus, Melanocetus Murrayi, Typhlonus nasus,* and *Aphyonus gelatinosus,* but not even a rudiment of an eye is to be found in *Ipnops Murrayi*.

But the fish of the greatest depths are by no means always characterised by small eyes. *Malacosteus,* a typical deep-sea form, has very large eyes, and so have *Bathylagus,* living in the enormous depth of 3,000 fathoms, and *Bathytroctes,* in 1,090 and 2,150 fathoms.

The result of recent deep-sea work, then, has been to show that as we proceed from shallow shore water to depths of 500 to 900 fathoms the eyes of the fish become larger, but in greater depths than 1,000 fathoms the eyes of some fish become considerably reduced, but those of others become still more enlarged. In the greatest depths of the ocean in fact it seems very probable that nearly all the fish are characterised by either very large eyes or very small ones.

We cannot expect to learn very much at present from the study of the eyes of deep-sea mollusca. The Cephalopods form the only class of this Phylum whose genera invariably possess large and well-developed eyes, and there does not seem to be any very marked increase or decrease in the size of the eyes of the few deep-sea cuttlefish that are known to us.

The eye of *Nautilus* is certainly remarkably interesting, but as this genus is the only representative of its order, and is known at times to float upon the surface of the ocean, it would certainly be erroneous to attribute the peculiarity of the structure of its eye to its 'temporary' deep-sea habits. We are still ignorant of the usual habitat of the remarkable genus *Spirula*, notwithstanding the fact that many of the tropical beaches are very largely composed of its empty shells. Whether it is a deep-sea dweller or not, we know nothing at present of the character of its eye, so that it can throw no light upon the problems we are now discussing.

Among the deep-sea gasteropods we find the same irregularity in the possession of eyes that we have just described among fishes. Thus a species of *Pleurotoma*, dredged by the 'Porcupine,' in 2,090 fathoms, has a pair of well-developed eyes on short footstalks, but *Pleurotoma nivalis*, obtained by the 'Talisman,' is blind. Again a species of *Fusus*, obtained by the 'Porcupine,' in 1,207 fathoms, is provided with well-developed eyes, but *Fusus abyssorum*, obtained by the 'Talisman,' is blind. Among the Lamellibranchs there are very few genera that possess well-marked eyes. The genus *Pecten* is one of those that in shallow waters possess numerous highly complicated visual organs situated on the edge of the mantle. In the deep-sea species, *Pecten fragilis*, these eyes are wanting, but we have not sufficient evidence at present to enable us to assert that all the deep-sea species of this genus are blind.

Among the Crustacea there is a very general tendency to lose the eyes at a depth of a few hundred fathoms of water.

In *Ethusa granulata*, for example, the eyes disappear at 500 fathoms and the eye-stalks become

THE CHARACTERS OF THE DEEP-SEA FAUNA 73

firmly fixed, greater in length, and take the place of the rostrum which disappears. In some forms —such as *Thaumastocheles zaleuca* and *Willemoesia* —the eye-stalks themselves have completely disappeared.

In the deep-sea Isopoda some forms lose their eyes entirely, but *Bathynomus giganteus* possesses a pair of enormous eyes, each provided with 4,000 facets.

To illustrate the distribution of eyes in this group, we may take as an example the genus *Serolis*. All the species of this genus are provided with eyes except *Serolis antarctica*—a species that extends from 600 to 1,600 fathoms.

The eyes of all the deep-sea species are relatively larger than those of the shallow-water ones, except *Serolis gracilis*, whose eyes seem to be disappearing.

But these large eyes of the deep-sea species of *Serolis* are not capable of any greater perceptive power. In fact, the evidence of degeneration they show, both in minute structure and in the diminution of pigment, proves that they can be of very little use to these animals for perception (see Figs. 4 and 5).

This increase in size, accompanied by degeneration of structure, is just what we should expect to find in the eyes of deep-sea animals, and it is difficult to explain why it is that we do not find more examples of it.

If the animals that now live in the depths of the sea are descended from the shallow-water forms of bygone epochs, they must have passed through many

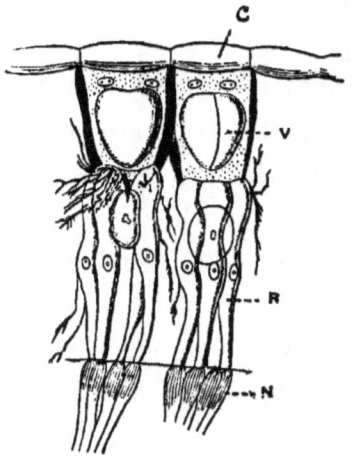

Fig. 4.—Semi-diagrammatic section through the eye of *Serolis schythei*, a shallow-water species (4–70 fathoms). C, lens; V, crystalline cone; R, rhabdom; N, nerve. (After Beddard.)

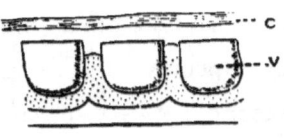

Fig. 5.—Diagrammatic section of the eye of *Serolis bromleyana*, a deep-sea species (100–1,975 fathoms), showing the degenerate character of the eye. The corneal facets C, and the crystalline cones V, are the only structures that can be recognised. (After Beddard.)

different habitats with diminished light until they reached their present dark abode in the abyss.

In every new region they came to, the forms with larger and better eyes would be at an advantage in the fainter light, and would be more likely to survive and transmit their favourable variation in this respect to their offspring, than their less fortunate neighbours. Thus down to the depth of the limit of sunlight we should expect to find, as we do find in fishes, large-eyed species.

Beyond the limit of direct sunlight the eyes would be of very little use to them, the pigment would disappear and the tissues become degenerate. This is precisely what has occurred in the genus *Serolis*.

The disappearance of the sense of sight in the animals of the deep sea is sometimes accompanied by an enormous development of tactile organs.

Thus, among fishes we find *Bathypterois*, a form that possesses extremely small eyes, provided with enormously long pectoral fin rays that most probably possess the functions of organs of touch.

Among the Crustacea we find the blind form, *Galathodes Antonii*, with an extraordinary development in length of the antennæ, and *Nematocarcinus*, with enormously long antennæ and legs.

The subject of the power of emitting phosphorescent light possessed by some deep-sea animals is much more difficult to deal with.

The presence of distinct organs in many of the deep-sea fish that can only be reasonably interpreted as phosphorescent organs, the presence of well-developed and evidently functional eyes in many deep-sea animals, and many other considerations render it very highly probable that some, if not many, forms emit a phosphorescent light.

The power and constancy of the light emitted, however, must for the present remain a matter of conjecture. We cannot judge at all of the amount of light given out by an animal in deep water by its appearance when thrown out of a dredge upon the deck. Whether the phosphorescent light given out by an Alcyonarian or a Crustacean is more or less at a temperature of 40° Fahr. and a pressure of one ton per square inch than it is at 60° Fahr. and the ordinary barometric pressure of the sea-level, is a question that has not yet been brought to an experimental test.

Whatever the answer to this question may be, the fact remains that a greater percentage of animals from the deep sea exhibit some sort of phosphorescent light when brought on deck than animals that live in shallow water.

The curious organs possessed by some fishes that are supposed to be organs for the emission of phosphorescent light have recently been subjected to a minute examination by von Lendenfeld.

It has been known for some years now, that the slime secreted by the skin glands of certain sharks is highly phosphorescent. It is not difficult, then, to understand how it came about that certain fish developed complicated phosphorescent organs.

From the phosphorescent slime secreted by a simple skin gland to the most complicated eye-like phosphorescent organ, we have a series of intermediate forms that are quite sufficient, even in the imperfect state of our knowledge at the present day, to enable us to understand the outlines of the evolution of these peculiar and interesting organs.

We can distinguish two kinds of phosphorescent organs in the deep-sea fish. There are the curious eye-like or ocellar organs situated usually in one or more rows down the sides of the fish's body, forming as it were a series of miniature bull's-eye lanterns to illuminate the surrounding sea (fig. 6); and various glandular organs that may be situated at the extremity of the barbels or in broad patches behind the eyes or in other prominent places on the head and shoulders.

Ocellar organs have been known for many years to occur on the sides of the interesting pelagic fish, *Scopelus*. Most of the species of this genus live in the open sea at moderate depths, coming to the sur-

78 THE FAUNA OF THE DEEP SEA

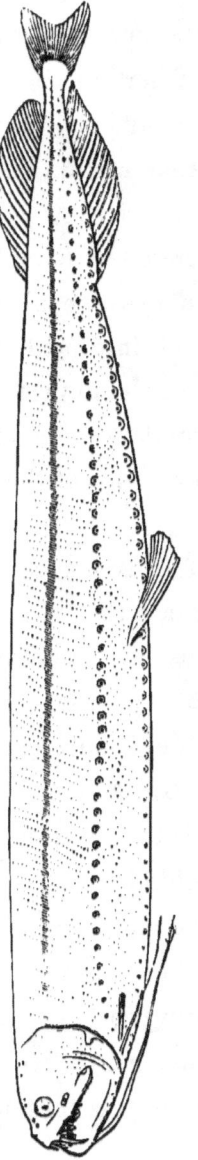

Fig. 6.—*Opostomias micripnus*; 2,150 fathoms. (After Günther.)

face only at night, but other species are found in almost every depth down to 2,000 fathoms of water.

In *Opostomias micripnus*, a dark black fish living at a depth of over 2,000 fathoms, there are two rows of ocellar organs running down the sides of the body from the head to the tail. In the living animal they are said to shine with a reddish lustre. In addition to these, the conspicuous organs, there are groups of fifty, a hundred, or even more very much smaller organs situated on the sides and back of the fish, each of which is lenticular in shape and consists of a number of short polygonal tubes containing a granular substance with rounded bases resting on the subjacent tissue. The whole organ is covered

THE CHARACTERS OF THE DEEP-SEA FAUNA 79

by a simple continuation of the cuticle of the body wall. The granular substance contained in the tubes is most probably the seat of luminosity.

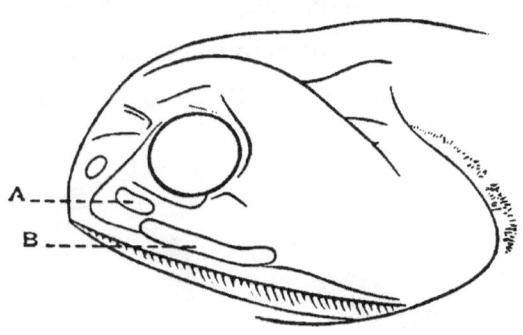

FIG. 7.—Head of *Pachystomias microdon* (after Von Lendenfeld). A, anterior sub-orbital phosphorescent organ; B, posterior sub-orbital phosphorescent organ.

As a type of the glandular organs we may take one of the sub-orbital organs found on the head of *Pachystomias microdon*.

In this fish there are two very conspicuous white organs immediately below the eye. The anterior one, which lies below and in front of the eye, is oval, with its upper margin slightly concave. In section it is seen to be surrounded by a thin layer of black pigment, and to consist of a reticular glandular substance in which is embedded a hammer-shaped lens-like body. Between these two structures there is interposed a thick layer of light reflecting spicules.

The exact part that is played by the different components of these curious phosphorescent organs is not yet known, but sufficient has been said to indicate to the reader the degree of complexity that these

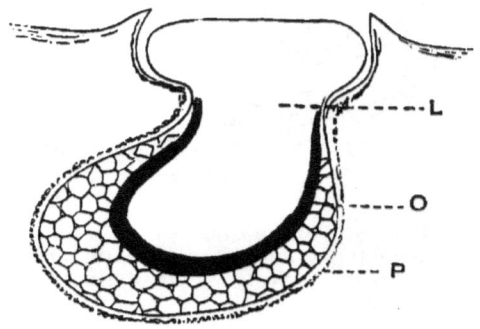

FIG. 8.—Section of the anterior sub-orbital phosphorescent organ of *Pachystomias microdon* (after Von Lendenfeld). L, lens; O, phosphorescent gland; P, pigment sheath.

organs may reach in the fish of the great depths of the ocean.

But the power of emitting phosphorescent light is by no means confined to the group of fishes. Some of the Macrurous Decapoda among the Crustacea are known to be phosphorescent. In the case of *Heterocarpus Alphonsi*, for example, the naturalists of the 'Investigator' found that 'clouds of a pale blue highly luminous substance, which not only illuminated the observers' hands and surrounding objects in the vessel in which the creature was confined, but also finally communicated a luminosity to the water itself, were poured out apparently from the bases of the antennæ.'

'The *Willemoesia*, too, was luminous at two circumscribed points somewhere near the orifices of the genital glands.'

Again, all the Alcyonarians dredged by the 'Challenger' in deep water were found to be brilliantly phosphorescent when brought to the surface, the light consisting, according to Moseley, of red. yellow, and green rays only.

Among the Echinoderms we have not many recorded instances of a phosphorescent light being emitted, but it is quite possible that many, if not all of them, may possess this power. The curious deep-sea form *Brisinga*, that was first discovered by Ch. Asbjörnsen, is known to be so brilliantly phosphorescent that it has been called a veritable *gloria maris*, and writing of the curious brittle-star *Ophiacantha spinulosa* (dredged by the 'Porcupine' in 584 fathoms of water), Professor Wyville Thomson remarks that the light was of a 'brilliant green, coruscating from the centre

G

of the disc, now along one arm, now along another, and sometimes vividly illuminating the whole outline of the star-fish.'

According to Filhol many of the abysmal Annelid worms are in the habit of emitting a vivid phosphorescent light, and capable thereby of illuminating the medium in which they live.

We have now considered all those characters exhibited by deep-sea animals that may be associated with the absence of direct sunlight. To run through them again briefly we may say: that the deep-sea species, belonging to classes of animals that usually possess eyes, show some modification in the size of their eyes, in that they are either very large, very small, or altogether wanting. That deep-sea animals are nearly always uniformly coloured. Very frequently they are black or grey or white, less frequently bright red, purple, or blue. But whatever the colour may be, spots, stripes, bands, and other markings of the body are very rarely seen. That deep-sea animals are brilliantly phosphorescent, the light being emitted either by special organs locally situated on the head, body, or appendages, or by the general surface of the body.

But there are some other characters that cannot be thus associated with the absence of sunlight.

In the first place bathybial fish, mollusca, crustacea, and other animals usually possess a remarkably small amount of lime in their bones and shells.

In fishes we are told that the bones have a fibrous, fissured, and cavernous texture, are light, with scarcely any calcareous matter, so that the point of a fine needle will readily penetrate them without breaking. In some the primordial cartilage is persistent in a degree rarely met with in surface fishes, and the membrane bones remain more or less membranous or are reduced in extent, like the operculum, which is frequently too small to cover the gills.

This cannot be due in all cases to a deficiency of carbonate of lime in the sea water, for we find these characters well marked in some of the fish, such as *Melanocetus Murrayi*, *Chiasmodus niger*, and *Osmodus Lowii*, that are found on the Globigerina mud.

Then again, the shells of the deep-sea Lamellibranchs, Gasteropods, Brachiopods, and Crustacea are very frequently remarkably thin and transparent, a character that is probably more generally due to a weakness in absorptive or secretive activity than to a deficiency in the supply of lime.

There are one or two characters of the deep-sea

fish that it is not easy to account for, and it is necessary only to mention their occurrence without attempting to offer any explanation of them.

One of the most common of these is the very dark pigment occurring in certain parts of the epithelium of the mouth and respiratory passages and the endothelium of the peritoneum. For example, in *Bathysaurus mollis*, living at a depth of 2,000 fathoms, the mouth and buccal cavities are black. The same thing occurs in *Ipnops Murrayi*, and indeed in all the strictly deep-sea forms.

Another important character of very frequent occurrence is the reduction in size, length, and number of the gill laminæ.

Among invertebrates we may mention as a fact of some interest, dependent perhaps on the soft character of the bottom, the preponderance of stalked forms over those of more sessile habits.

Thus among the Alcyonaria the characteristic forms of the deep water are the Pennatulids, and more particularly the genus *Umbellula* with its long graceful stem and terminal tuft of polyps. Among the Echinoderma we find many forms of stalked Crinoids. Among the Tunicates several curious genera characterised by their long peduncles.

Taking the fauna as a whole, Moseley regarded it as similar in some respects to the flora of the high mountains. Some forms are dwarfed in size, such as the species of Radiolaria, Cerianthus, some of the Cephalopods, &c., while others are very much larger than their shallow-water allies, such as the Pycnogonids, nearly all the Crustacea, Alcyonarians (as regards the size of the polypes), Siphonophora, and many others.

CHAPTER V

THE PROTOZOA, CŒLENTERA, AND ECHINODERMA OF THE DEEP SEA

The most important, but perhaps somewhat disappointing, result of the deep-sea researches of recent years has been to prove that the abysmal fauna does not possess many very extraordinary forms.

It seemed probable, before the dispatch of the 'Challenger' expedition, that when the dredge and the trawl should be successfully employed in depths of over 2,000 fathoms, a new and remarkable fauna would be brought to light. Some naturalists thought it even possible that, not only would many genera be found alive that are known to us only by their fossilised skeletons in the secondary and tertiary rocks, but that there might be many other new creatures whose anatomy would throw much light on the theories of the evolution of the animal series.

But none of the great expeditions that have sailed since the year 1874 have yet succeeded in showing that the hopes and wishes of these naturalists were really justified. Although thousands of species of animals have been described in the volumes that have been devoted to deep-sea work, the number of the sub-kingdoms and classes remains the same, and indeed the number of new families and genera has not been increased in any very unprecedented manner.

We have found no animals in the depths of the sea of such interest and importance as Ornithorhynchus, Amphioxus, Balanoglossus, Peripatus, Millepora, or Volvox among the living, or Hipparion, Archæopteryx, Ammonites, Slimonia, and the Trilobites among extinct animals.

The abysmal fauna is not in fact remarkable for possessing a large number of primitive or archaic forms. It is mainly composed of a number of species belonging to the families and genera of our shallow-water fauna that have, from time to time, migrated into greater depths and become modified in their structure in accordance with the extraordinary conditions of their new habitat.

There is very good reason to believe that this migration has been going on from time immemorial, and consequently we find a few forms typical of the bygone times, left to struggle for existence with the more recent immigrants from shallow waters. But after all the proportion of ancient forms to modern ones in the fauna of the abyss is not larger than it is in the fauna of fresh-water lakes and streams or even of the dry land. Nor is there any reason why it should be. The land and the fresh water have been peopled by migrations from the shallow water of the sea from generation to generation in precisely the same way, and they each can show a certain number of archaic forms.

We must now consider briefly some of the most interesting deep-sea representatives of the various classes of the animal kingdom, referring as we pass on to the extent to which these classes contribute to the fauna of the abyss.

We find a great difficulty in determining with any degree of certainty the actual depths at which the supposed abysmal forms of Protozoa actually live. All the Radiolaria and Foraminifera—the only Protozoa that are largely represented in the fauna of the open seas—are planktonic in habit; that is to say, they float or drift about in the water without ever becoming attached to the sea bottom; and when the contents of a dredge, that has been hauled up from a great depth, are examined, it is impossible to say at what points in its long journey from the bottom the Protozoa it contains were caught. Even if dredges and nets are used which can be closed by a messenger at any particular depth, the problem cannot be very easily settled; for even if the protozoa shells that are captured are found to contain a certain amount of protoplasm, it must be proved that that protoplasm is actually alive when brought on deck before we know for certain that the species actually live on the bottom. When the pelagic Foraminifera and Radiolaria die and sink to the bottom, their protoplasm probably disintegrates very slowly, and it is quite probable that the floor of the ocean is littered with the shells of truly pelagic protozoa, each containing a greater or smaller amount of undecomposed protoplasm.

However, there is little doubt that there are some truly abysmal Protozoa. Among the Radiolaria, for example, it seems extremely probable that the majority of the Phæodaria and many Spumellaria live only in very deep water. 'A character common to these abyssal forms,' says Haeckel, ' and not found in those from the surface or slight depths, is found in their small size and massive heavy skeletons, in which respects they strikingly resemble the fossil Radiolaria of Barbadoes and Nicobar islands.' The Phæodaria are very widely distributed over the floor of the ocean, and occur in some districts in such numbers that the 'Challenger' was able to bring home some hundreds of thousands of specimens. They are distinguished from other Radiolaria by the thick outer and thin inner capsule, by the typical main opening or atropyle placed on the oral pole of the main axis with a radiate operculum provided with a tubular proboscis, and lastly by the presence of the phæodium, a voluminous pigment body which lies invariably on the oral half of the calymma and is composed of numerous singular pigment granules of green, olive, brown, or black colour.

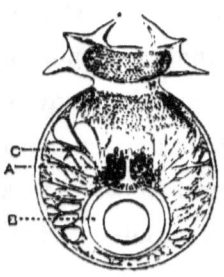

IG. 9. — *Challengeria Murrayi*, one of the Phæodaria (2,250 fathoms). A, phæodium; B, central capsule; C, strands of protoplasm in the calymma. After Haeckel.

There are many genera belonging to the Foraminifera that are very probably inhabitants of abysmal depths, but they do not seem to possess any special characters, unless it be a greater thickness and density of their shells, to distinguish them from their shallow-water allies.

Passing now to the group of the sponges or Porifera, we find that the calcareous sponges are not represented at all in the abysmal zone. Two species are found at a depth of 450 fathoms, but none are truly bathybial in habit. The same remark applies to the horny sponges. These forms chiefly belong to the littoral or very shallow-water fauna, and never descend to greater depths than 100 fathoms. Of the other groups of Porifera—the Monaxonia, the Tetractinellidæ, and the Hexactinellidæ—several genera are known to extend down to some of the greatest depths at which trawling operations have been successfully carried on. It is difficult to point to any characters in these sponges that can be attributed in any way to the conditions of deep-sea life, but nevertheless we do find in deep water some of the most remarkable and beautiful forms of sponge skeleton that can be found anywhere.

Amongst the Cœlentera we find in the deep water a remarkable sub-family of Medusæ, which has been named by Haeckel the Pectyllidæ. It is distinguished from the other jelly-fish by the curious sucking cups situated at the ends of the tentacles. It seems probable that they are used for purposes of locomotion, the animal walking over the muddy bottom as on a series of stilts.

Like most of the deep-sea Hydroids, the Pectyllidæ are usually devoid of sense organs, but a single specimen of *Periphylla mirabilis*, captured by the naturalists of the 'Challenger,' possessed well-marked eyes.

There is also a peculiar family of the Siphonophora, called the Auronectæ, consisting of a few specimens that have been hitherto found only in very deep water. Like the well-known Portuguese man-of-war *Physalia* of the surface waters, the Auronectæ possess a large swimming bladder or pneumatophore, but they have in addition another peculiar bladder-like cavity, called the aurophore, communicating with it, which may be an organ for secreting gas.

A very interesting genus allied to Velella was also found in depths of over 2,000 fathoms by the 'Challenger' expedition. It is supposed to be a survival of the ancestral form of the Disconectæ, or, at any rate, to be a link connecting the Siphonophora with the Medusæ. The very well marked octoradial arrangement of the parts of *Discalia*, as this genus has been termed, is certainly a point of great interest and importance.

There is no large family of the sea anemones that is peculiar to deep water, but several genera that occur only in the abyss exhibit some curious modifications. The manner in which the tentacular pores have become enlarged, and the tentacles themselves diminished in size and flexibility, has already been referred to in a previous chapter (p. 36).

The family of sea anemones that has been named the Corallimorphidæ, characterised by the stiffness and slight contractility of the body, the knobbed nature of the tentacles, and their distribution in several series, was, until quite recently, considered to be a true abysmal family. The two species, *Corallimorphus rigidus* and *C. profundus*, are known to occur only in very deep waters, and present some curious modifications of structure in relation to their habit; but it seems probable that to this family should be added the remarkable littoral form *Theluceros rhizophoræ* found on the coast of Celebes attached to the roots of the mangrove trees in the swamps.

The fact that all the principal groups of the Actiniaria, except, perhaps, the group that includes those forms with only eight mesenteries, the Edwardsiæ, have representative genera or species in the great depths of the ocean, points to the conclusion that the sea anemones have migrated from the shallow waters in comparatively recent times, and that the migrations have been successive, each period of their history sending some specimens to survive or to become extinct in the struggle for life in the deep sea.

Of the Madreporarian corals, several genera are now known to inhabit very deep water, but they do not present many very remarkable points of divergence from the shallow-water forms.

It is true that as we pass from the shallow waters, of those parts of the world where the great colonial madrepores build up the greater part of the vast coral reefs, into the deeper water beyond them, the solitary forms become relatively more abundant, but no new groups characterised by any special deep-sea attributes make their appearance. We must remember, not only that a great many solitary corals occur in shallow water in different parts of the world, but that some colonial forms, such as *Lophohelia prolifera* for example, are found only in very deep water.

Until quite recently it was usually stated in works dealing with the structures of coral reefs that the so-called reef-building corals, that is to say the large madrepores, astræids, and others, are confined to water not deeper than thirty fathoms. This limit must now be somewhat extended, in consequence of the discovery by Captain Moore of an abundance of growing coral at a depth of forty-four fathoms in the China seas ; but, nevertheless, it is perfectly true that the corals do not grow in such profusion in very deep water as to form anything that can be compared with the reefs of the shores. It is quite possible that the advantages afforded by the light, warmth, and abundance of food of the shallow water may account for the luxuriance and vigour of the reef corals, and that where the food is scarce, and the water cold and dark as it is below fifty fathoms, the power of continuous gemmation is lost, and the rapidity of the growth and reproduction of the individual polyps is considerably diminished.

The fact remains, however, that, as with the sea anemones, so with the madrepores, nearly all the great divisions have a few isolated representatives in the abyss, and that no great family occurring in large numbers has yet been discovered peculiar to this zone.

The Alcyonaria, on the other hand, do present us with at least one example of a true deep-sea family. This great class of Anthozoa, distinguished from the Zoantharia by the presence of not more than eight tentacles and mesenteries and by the pinnate character of the former, falls into four principal divisions. The Stolonifera, the Alcyonidæ, the Gorgonidæ, and the

Pennatulidæ. The first three of these divisions principally inhabit the shallow water. Each of them sends a few representatives into the great depths, but by far the greater number of the genera and species are to be found between tide-marks or in depths of less than fifty fathoms.

The Pennatulids, on the other hand, are rarely found in very shallow water, and nearly half the known genera live in deep water. At least two families may be said to be characteristically abysmal. These are the Umbellulidæ and the Protoptilidæ.

The Pennatulidæ are regarded by naturalists as the most complicated or highly organised group of the Alcyonaria. Three different forms of polype build up the colony or sea-pen as it is called. There is a single very much modified and enormously large polype, without tentacles, forming the axis, a large number of ordinary Alcyonarian polypes (autozooids) arranged in the form of leaves, or simply scattered irregularly on the surface of the central polype, and a number of very small undeveloped polypes (Siphonozoids) without tentacles, whose function seems to be to pump water into the canals of the colony, and thus to keep up the circulation of water.

The deep-sea genus *Umbellula* possesses a very

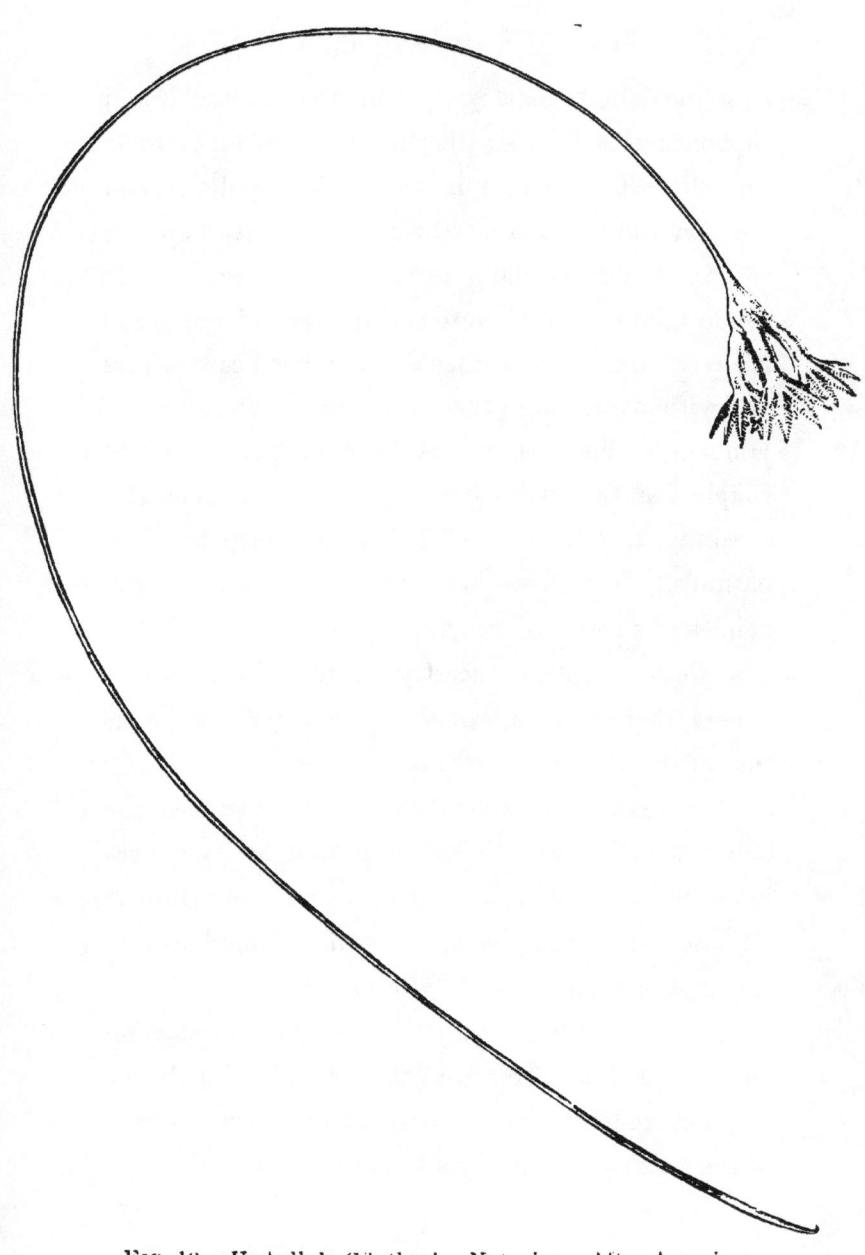

FIG. 10.—Umbellula Güntheri. Nat. size. After Agassiz.

long and delicate axial polype, and the Autozooids and Siphonozooids form a little cluster only at its extreme summit. The small number of these polypes and the very limited area over which they extend are the two most characteristic features of the genus. It would take me too far into the anatomy of the group if I were to add any further details; but I cannot pass on without noting that the whole structure of Umbellula shows that it is far more primitive and simple than the shallow-water genera. And, generally speaking, this holds good for all the deep-sea Pennatulids. In fact, we have here one of the rare examples of a series of genera, that can be regarded as a slightly modified ancestry of the shallow-water genera, that has been brought to light by the exploration of the abysmal depths of the ocean.

We have seen, then, that of the Cœlentera, the only order that has a large proportion of its genera living in deep water, is the only one whose members all possess a stalk by which they fix themselves into the mud or sand at the bottom of the sea.

It is not uninteresting to note, then, in passing on to the Echinoderma, that the stalked Crinoids, the only Echinoderms that can permanently fix themselves to the bottom, are nearly all found in deep water.

Several years before the 'Lightning' was despatched on her memorable pioneering voyage, Vaughan Thomson had proved that the common feather star of the shallow waters of the British coasts passes through a stage in its development which resembles the fossil genera of the order in being provided with a stalk for attachment.

But it was left for the naturalists of the 'Porcupine,' the 'Challenger,' the 'Talisman,' and other vessels employed in deep-sea researches to prove that adult stalked Crinoids are still living in nearly all parts of the world at the great depths of the sea.

The genera of stalked Crinoids now living are the remains of a family that at one time had many representatives in all parts of the world. Nearly all the marine deposits of bygone epochs, including even those of such remote periods as the Cambrian and Sub-Silurian, contain the fossilised skeletons of these Crinoids. In some strata they are represented by only a few genera, but in others they are found in such enormous numbers that the sea-beds of those early times must have been literally carpeted with them.

At the present day the few genera that survive have been driven from the shore waters, and are

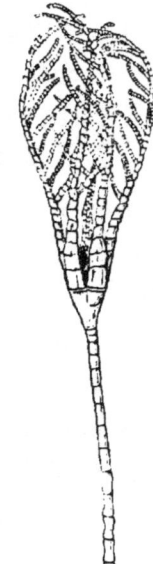

FIG. 11.—*Rhizocrinus lofotensis*, one of the deep-sea stalked Crinoids. (After Carpenter.)

chiefly found at depths of more than 200 fathoms, a few only extending into 140 and even 70 fathoms.

There are six genera known, and of these, two, *Hyocrinus* and *Bathycrinus*, are not found in less than 1,000 fathoms of water.

There can be no doubt that these modern stalked Crinoids are closely related to many of those that flourished in bygone periods of the history of the earth. As Carpenter has pointed out, the family Pentacrinidæ are remarkable for their long geological history. The genus *Pentacrinus* itself first appeared in the Trias and persisted through the Secondary and Tertiary times to the present day.

The general character of the fossil Pentacrinidæ is essentially the same as that of their recent representatives, except that they often had much longer stems which reached to a length of as much as 50 or even 70 feet; while the number of arms was frequently limited to ten, which is not the case in any recent species but *Pentacrinus naresianus*.

But the deep-sea Echinoids, or sea-urchins, also present some features of particular interest. Professor Agassiz in his report says, 'One of the very first results clearly indicated by the deep-sea dredgings of Count Pourtales and the subsequent investigations of the " Porcupine " expedition was the antique character of the new genera discovered in deep water, and especially their resemblance to the cretaceous genera; and the study of the " Challenger " Echinoids has brought this out still more clearly.'

No fewer than twenty-four genera extend into the abysmal regions; of these no less than sixteen, nearly all belonging to a new group of Spatangoids, do not live at all in shallow water.

The most interesting forms among these are the Pourtalesiæ, a group that has existed since the Chalk. These are heart-shaped urchins with a very peculiar test. 'They all have large coronal plates, recalling the Echini, with a disconnected apical system characteristic of many cainozoic spatangoids; they have a sunken anal system, some of them a most remarkable anal beak and a very striking pouch in which the mouth is placed.' They are found only in very deep water, and have no allies among the modern littoral fauna.

The genera *Calveria* and *Phormosoma* are two of the most abundant Echinoids found in deep water, and they are both representatives of forms that were very abundant in cretaceous times. They are remarkable for the extreme flexibility of their shells. In shallow-water sea-urchins the shells are composed of a great number of little plates that fit so closely to one another that no movement is possible between them. When the animal dies all the soft tissues decay and the shell remains, to be tossed about by the waves until crunched or dashed to pieces. In *Phormosoma*, however, the tiny plates of which the shell is composed are freely movable on one another, and when the animal is alive very considerable contractions and expansions can take place.

None of the modern shallow-water Echinoids present this peculiarity, and it is a very interesting and surprising fact that in this respect the fossils of the chalk should resemble so closely the living urchins of the abyss.

But before leaving the Echinoids reference must be made to two more points that have been made by the illustrious American naturalist. Agassiz points out that all those genera that have the greatest bathymetrical range, extending from the littoral to the abysmal region, are at the same time genera which date back to the Cretaceous period, while those having a somewhat more limited range go back to the tertiaries, and those that extend only slightly beyond the littoral area go back only to the later tertiaries.

This interesting generalisation brings home to our minds the enormous length of time that it must have taken these animals to migrate from the shallow to the deep sea. In the struggles for existence between marine animals it must always have been the last resort of those unable to compete with the younger generations in shallow water to migrate into the deeps.

The scarcity of food, the darkness, and the pressure of these regions can never be so favourable for the support of animals as the conditions of the shores. We can well imagine that a species would take every opportunity that is afforded to return from such inhospitable habitats, and that only when, as it were, every door is closed, when no island, continent, or cape can afford it a free scope for life in shallow water, does it become a true deep-sea species.

Steps taken towards the darkness in one period may be retrieved in the next. The competing species may itself have become extinct or have moved to another locality. Organs may have become modified or a new source of food supply tapped which enable them to return once more to shallower waters. No wonder that the steps in the progress, or rather retreat, to the abyss have been the work of a time that can be counted only by geological periods; and no wonder then at the remark made by many deep-sea naturalists that the abysmal fauna becomes poorer the farther it is from shallow water.

The group of the Asteroidea, or star-fishes, contributes largely to the fauna of the abyss.

During the voyage of the 'Challenger' no fewer than 109 different species were found in depths of over 500 fathoms, and in some localities a very large number of star-fish were taken in one haul of the dredge.

Nevertheless, there are not many abysmal genera that differ to any remarkable degree from the littoral ones; and indeed it may be said that the recent work on deep-sea Asteroids does not throw much new light either on the phylogeny of the group or on their palæontological history.

The genus *Brisinga*, at one time supposed to be a connecting link between the star-fishes and the brittle stars (Ophiurids), has recently been shown to be closely related to the families Heliasteridea, Echinasteridea, and others typical of the class Asteroidea; and, as Sladen has pointed out, the peculiarities of structure that it exhibits are probably due to its extreme isolation and the influence of its abysmal habitats.

But no work on the deep-sea fauna would be complete without some reference to Brisinga. Discovered by Asbjörnsen in 1853, in 200 fathoms of water in the Hardanger fjord, and described in a splendid memoir by the elder Sars, it excited great interest among naturalists. The great brilliancy of the phosphorescent light that it gave out on being brought on deck, the remarkable tendency that it had to cast off some of its numerous long, thin, ophiurid-like arms, and some of the general features of its internal anatomy were points that were considered at the time to be sufficient to justify the establishment of a separate sub-order for the family Brisingidæ.

The more recent discovery, however, of genera allied to Brisinga has bridged over the gap separating it from other star-fish, and it is now considered simply as the type of a family of the order.

The numerous species of the genus that have been found since Asbjörnsen's original discovery are all inhabitants of deep water, some of them going down to the enormous depth of 2,000 fathoms; indeed there are very few genera in the animal kingdom, containing so many species as the genus Brisinga, that have such a uniform deep-sea habitat.

The last group of Echinoderms that we have to consider is the Holothurians. It contains one order—the Elasipoda—that may be considered to be truly bathybial, as there is only one species belonging to it, *Elpidia glacialis*, that extends into water as shallow as fifty fathoms.

The Elasipoda are remarkable for their strongly-developed bilateral symmetry. Adult Echinoderms as a rule possess a well-marked radial symmetry, as we see exemplified in the feather-star, star-fish, and sea-urchin, but this radial symmetry is only adopted when they undergo their metamorphosis from the free swimming and bilaterally symmetrical larval stage

They are not born radially symmetrical, but become so as they grow up. Moreover, we must bear in mind that the radial symmetry of the adult only obscures, it does not obliterate, the bilateral symmetry of the larva.

In the Holothurian, however, we can always discover a clear bilateral symmetry even in the adult. That is to say, we can recognise an anterior and a posterior end, a right and a left side of the body. It is an organisation which emphasises, as it were, the anterior and posterior ends, the right and left sides and the dorsal and ventral surfaces that characterise this interesting deep-sea order, the Elasipoda.

Here, then, we have an example of a character common to all the larvæ of the sub-kingdom and exceptionally well marked in the adults of a family confined to deep-sea habitats.

Now we know that there is a tendency for some of the peculiar characters of the ancestors of animals to be recapitulated in the course of their development from the egg, and accordingly most naturalists are agreed that all the Echinoderms have descended from some form of bilaterally symmetrical ancestor. Are we, then, to believe that the Elasipoda brought from the depths of the sea are more closely related to these ancestral forms than the shallow-water families?

The state of our knowledge at the present day hardly allows us to answer this question very definitely. However nearly they are related to such ancestral Echinoderms in general form, they are probably profoundly modified by a deep-sea life. Nevertheless, in the simple shape of the calcareous corpuscles of the skin, the simple form of the calcareous ring, the communication of the water-vascular system with the exterior by one or several pores, and in some other anatomical characters, they give evidence of their primitive characters.

CHAPTER VI

THE VERMES AND MOLLUSCA OF THE DEEP SEA

It has not been my intention in this volume to confine my attention to the truly abysmal forms, but rather to consider all those animals living in deep water that show any characters strikingly different from their relatives living in shallow water.

The term deep water is, after all, only a relative one.

To one accustomed only to shore collecting, ten fathoms is deep water, while on the other hand, to such naturalists as those on board the 'Challenger,' who are accustomed to dredge in all seas, nothing under 1,000 fathoms is considered deep water.

We must bear in mind, however, that at a depth of only 200 fathoms, the conditions of life are very different to those of the shore waters. We find a very great diminution in the amount of light, for instance, that can penetrate through sea water teeming with floating organisms of all kinds to reach the fauna attached to the bottom at such a depth. The diminution in the amount of light must mean a diminution in the rapidity of growth of chlorophyll-bearing plants, and consequently a diminution in the food supplies of animals drawn from that source.

We might expect then to find, even in such shallow water as this, some forms of particular interest. It is true that the greater part of the fauna is made up of ordinary shallow-water forms that have migrated quite recently, and perhaps only temporarily, into the depths, but we expect to find, and actually do find, the outposts of a new fauna.

These remarks lead me to the consideration of one or two very remarkable animals that have recently been brought to light.

In that strange assembly of animals which, for want of a better word, the authorities call the Vermes, there are three groups whose relations to one another and to the other groups of Vermes have been and still remain a puzzle to naturalists.

These three groups are the Gephyrea, the Polyzoa, and the Brachiopoda. In external form they are as different from one another as possible.

The Gephyrea are solitary worm-like forms burrowing in the sand or perforating rocks; the Polyzoa are minute creatures that frequently build up by budding large colonies which assume in some cases dendritic forms like corals, and the Brachiopoda are protected by thick bivalve shells simulating in a striking manner the shells of the Lamellibranchiate mollusca.

But external form is not the only character that can be relied upon for purposes of classification. The general and minute anatomy, together with the story of the development of these animals, teach us that they are in some way closely related.

It is not within the scope of this book to enter into the discussion of what these relations are; suffice it to say that the controversy has within recent years to a great extent turned upon the position in our classification of three interesting genera. These are Phoronis, Rhabdopleura and Cephalodiscus.

Phoronis occurs only in shallow water, Rhabdopleura has been found in water from 40 to 200 fathoms deep off the Shetlands and on the Norwegian coasts, while Cephalodiscus was discovered by the 'Challenger' at a depth of 245 fathoms off Magellan Straits.

Rhabdopleura forms colonies consisting of branched tubes growing upon the tests of Ascidians,

112   THE FAUNA OF THE DEEP SEA

on sea-weeds, corals, or other objects fixed to the sea-bottom. In the open, free extremity of each of the

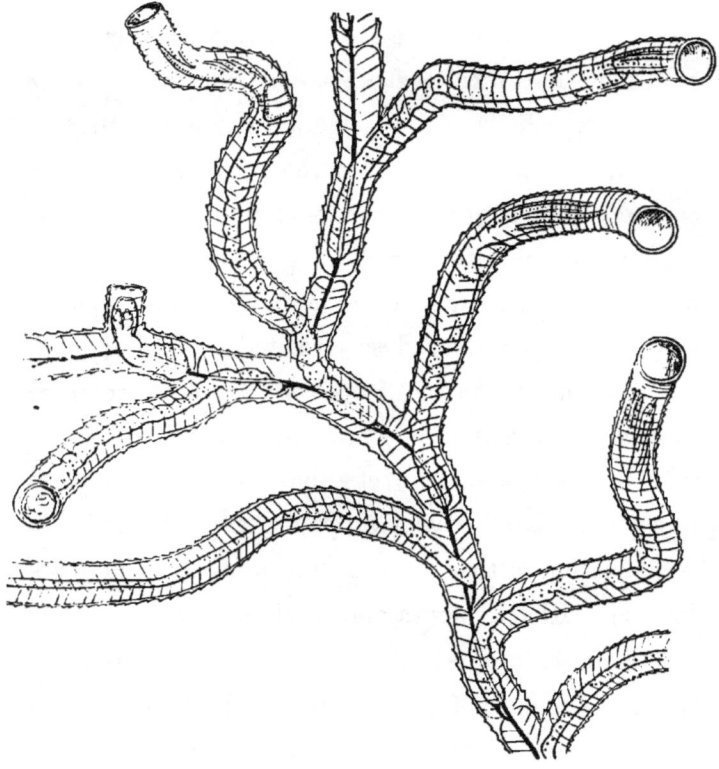

FIG. 12.— A portion of a colony of *Rhabdopleura normani*. (After Lankester.)

branches may be found the polypide attached to a filament or stalk which connects it with the other polypides of the colony (fig. 12).

Each polypide is provided with a single pair of large pinnate arms, resembling the arms of a Brachiopod, and a broad muscular epistome by means of which it is able to creep up or down the tube.

The affinities of this interesting creature are by no means sufficiently well understood. It is one of those forms that, without being, strictly speaking, a connecting link between large and well-known groups of animals, indicates to us some of the lines of evolution that these groups may have passed through; and, in so far as it does this, it has its value and importance.

Cephalodiscus, though related to Rhabdopleura in the presence of a structure corresponding to the arms, and a broad epistome, seems to be more closely connected with such a form as Balanoglossus in the presence of a single pair of gill-slits, a small rudimentary notochord and the position of the central nervous system.[1]

Whatever position these genera may ultimately occupy in our systems of classification, there can be little doubt that much valuable information will be obtained by a further study of their structure and

[1] A rudimentary notochord projecting forward from the buccal cavity into the epistome has quite recently been discovered in Rhabdopleura.

I

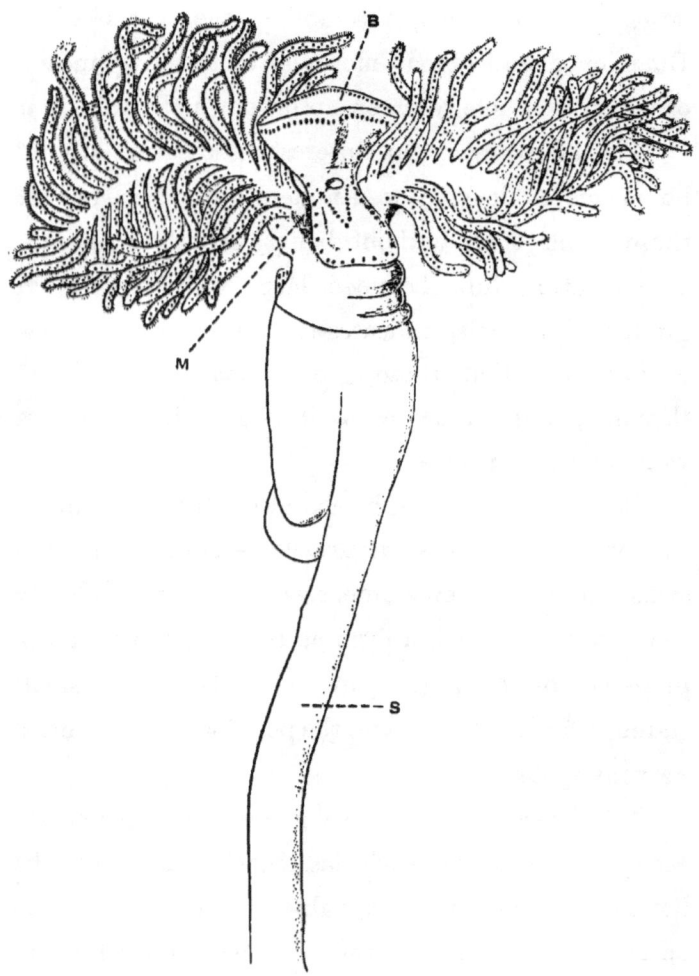

Fig. 13.—A single polypide of *Rhabdopleura normani*. M, mouth; B, epistome; S, polypide stalk. (After Lankester.)

development—information that will probably shed much light on the relationships to one another of the many groups of Vermes. Their occurrence in water of moderate depths only indicates perhaps that they are gradually being crowded out from the more favourable localities of shallow water, and are tending towards extinction on the one hand, or a deep-sea habitat on the other.

The Brachiopoda need not detain us long. Some species are capable of existing at a great variety of depths without any observable modification of shape or characters. Thus *Terebratulina caput serpentis* has the extraordinary bathymetrical distribution of 0–1,180 fathoms, and *Terebratula vitrea* 5–1,456 fathoms. *Atretia* is the only genus peculiar to deep water. It is a noteworthy fact in connection with this order that the two genera, *Lingula* and *Glottidia*, which compose the sub-order Ecardines, are both confined to shallow water. Now the Ecardines are anatomically, at any rate, the most primitive of the Brachiopoda, and Lingula has the most ancient geological history of any living genus of the animal kingdom, shells almost identical with those of the living species being found abundantly in the Cambrian strata. Why it is that Lingula has been able to maintain itself almost unchanged through all the countless generations that have elapsed since Cambrian times, and can now flourish amid the desperate struggle for existence in the shallow waters of the tropics, while its companions, the corals, mollusks, arthropods, &c., have changed or passed away, is one of those problems in natural history that seems to us impossible of solution. The time may come when we shall be able to appreciate better than we do now the complicated relations between animals and their environment, and then perhaps the peculiar fitness of Lingula will be made manifest; but at present we can but mention the fact as a fact, and leave the solution of the problem to the future.

The order Gephyrea is probably another very ancient group of animals, although in the absence of any hard calcareous, siliceous, or horny skeleton the geological record can give us no confirmation of their antiquity. As with the Brachiopods so with the Gephyrea, some of the species have a very wide bathymetrical distribution. *Sipunculus nudus*, for example, the commonest and best known of all the Gephyrea, extends from quite shallow water to a depth of over 1,500 fathoms. As Selenka has pointed out, it is those Gephyreans that live in holes in stones, or in shells such as *Phascolion* and *Phascolosoma*, that are more frequently found at the greater depths; but, apart from this, there are no characters that exclusively belong to the abysmal Gephyrea or are more frequently found in them than in the shallow-water forms. Nor are there any genera, at present brought to light, that are confined to those regions of the sea.

The group of the Annelida is not very well represented in the deep-sea fauna. The genera *Serpula* and *Terebella* have been found very widely distributed over the earth, at all depths from the shore to the abyss, but there do not seem to be many genera that are confined to deep water. In some cases, where there is a scarcity of lime in the water, the thin protecting tubes of the sedentary forms are strengthened by the adhesion of foreign particles, such as sponge spicules and arenaceous foraminifera, but in others, the tubes are formed of successive layers of a transparent quill-like substance (*Nothria Willemoesii*) which is frequently armed with spiny projections.

Most of the errant Polychætes found at great depths are said to be most brilliantly coloured, and some of these, such as *Eunice amphiheliæ*, have the power of emitting a bright phosphorescent light; but there seem to be no very definite and constant characters separating these forms from the Polychætes of shallow waters.

As is the case with many other orders of animals, the species of Annelida living in deep water are either blind or possess eyes of a remarkably large size. *Genityllis oculata* may be taken as an example of a deep-sea annelid with large eyes. This annelid, belonging to the family Phyllodocidæ, was found at a depth of 500 fathoms in the Celebes sea. It possesses two enormous eyes which cover almost the whole of the head, and there can be no doubt, from the investigations of Dr. Gunn on their minute anatomy, that they are perfectly functional.

Before leaving the Annelida a brief notice must be made of the very extraordinary form *Syllis ramosa*, found parasitic on a hexactinellid sponge at depths of about 100 fathoms. It is chiefly remarkable for the very complicated manner it has of producing buds which do not immediately become detached from the parent, but form a compound network which ramifies through the interstices of the sponge like the colony of a Hydromedusan.

Passing now to the sub-kingdom Mollusca, we shall find that all the classes are represented in the abysmal fauna.

The Lamellibranchiata, or bivalves, occur in almost all depths of the ocean, *Callocardia pacifica* and *Callocardia atlantica* having been found at the enormous depth of 2,900 fathoms. Some species, such as *Venus mesodesma*, have a very wide bathymetrical distribution, but others are only known to occur in deep water.

Concerning the characters of the deep-sea Lamellibranchiates, Mr. Smith, in his report on the Lamellibranchia of the 'Challenger' expedition, says 'very deep-water benthal species certainly have a tendency to be without colour, and of thin structure, no doubt resulting from the absence of light, the difficulty of secreting lime, the scarcity of food and other unfavourable conditions of existence.' But notwithstanding this, the same author continues: 'The species are apparently few in number in comparison with those of shallow water; and new and peculiar generic forms which we naturally expected would have been discovered are of even still rarer occurrence.'

As in the group of the Lamellibranchia, so in the Gasteropoda, no very remarkable new genera or species have been found in deep sea. Some shallow-water genera, such as *Fusus* for example, have representative species in the abyss; but, with the exception of a want of brilliant coloration and marking and a thinness of the shell, the deep-sea forms do not exhibit any characteristic features. One of the most brightly coloured shells found at great depths is that of *Scalaria mirifica*, which is tinted rose and white, but this seems to be quite an exceptional character among the deep-sea Gasteropods. Several new genera were found in the deep water, but their general characters do not call for any special remark.

Among the Cephalopoda there seems to be little doubt that the genera *Cirroteuthis*, *Bathyteuthis*, and *Mastigoteuthis* are entirely abysmal, and the same applies probably to one or two species of octopus; but as Hoyle remarks, ' apart from the single fact that *Bathyteuthis* and *Mastigoteuthis* both have slender filiform tentacles with minute suckers, no structural features have been discovered which will serve to diagnose a deep-sea form from a shallow-water one.'

The exact habitat of the interesting genus *Spirula* is unfortunately still unknown. In some parts of the tropics the shores are covered with spirula shells, and yet the animals that secrete them are still to be reckoned amongst the greatest rarities of our museums. The numerous dredgings of the 'Challenger' only brought to light one specimen of this animal, and that from a depth of 360 fathoms, and the

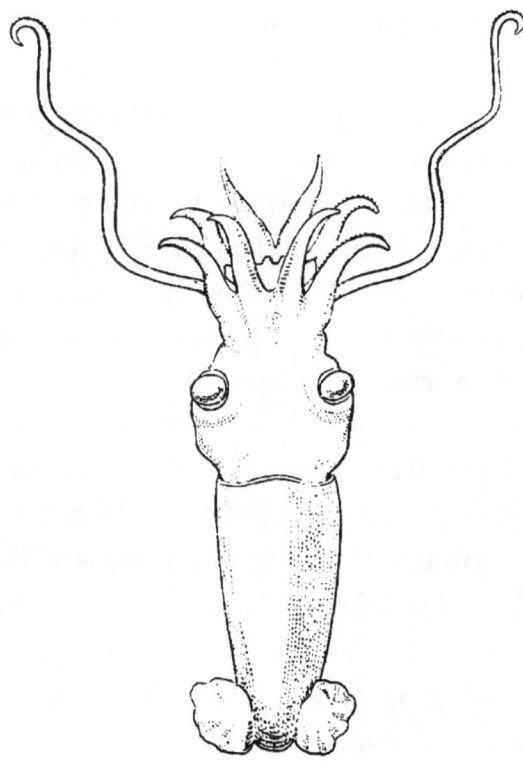

Fig. 14.—*Bathyteuthis abyssicola.* (After Hoyle.)

'Blake' caught one at a depth of 950 fathoms, so that there can be little doubt that *Spirula* lives in deep water.

It seems to be very probable that some day, when the right place and depth are discovered, *Spirula* may be discovered in great abundance, but we have at present no means of judging whether this will be in very deep water or not.

Almost precisely similar remarks apply to the distribution of the pearly Nautilus. The shells of this Cephalopod are sometimes found in great number on the shores of some of the islands of the Southern Pacific Ocean and the Malay Archipelago, but the living animal is but rarely captured. It has been asserted by some travellers that the pearly Nautilus floats on the surface of the ocean and possesses the power of suddenly diving to great depths on being disturbed; but it must be remembered that Rumphius originally caught his Amboyna specimens of Nautilus in traps set at a depth of 200 fathoms baited with sea-urchins, and that the 'Challenger' captured a single living specimen off Matuku island in 300 fathoms.

The probability, then, is that both *Nautilus* and *Spirula* should be included in the deep-sea fauna, but we are still in want of a great deal more information concerning their life and habits before this point can be definitely determined.

CHAPTER VII

THE ARTHROPODA OF THE DEEP SEA

THE deep-sea fauna seems to be particularly rich in marine Arthropoda, many curious and interesting forms being brought up with almost every haul of the dredge. The Arthropoda, too, being very highly organised animals, afford interesting and instructive examples of the effect of abysmal life in the modification of the sense organs and the production of varieties specially modified for the conditions of the struggle for existence in their strange habitat.

Concerning the groups of Ostracoda and Copepoda it may be said that the evidence is not yet conclusive that they include any truly deep-sea species. The largest known Ostracod, measuring somewhat more than an inch in length and probably allied to the genus *Crossophorus*, has quite recently been captured by Professor Agassiz at depths of less than 200 fathoms, but he could obtain no evidence that it descended into much deeper water than this.

Mr. Brady, in writing the report of the 'Challenger' Ostracoda, came to the conclusion that they do exist in very limited numbers in the most profound depths of the sea; but it is nevertheless quite possible that all the Ostracods brought on deck by the trawl or dredge were really captured either on the way down or on the way up, and are, strictly speaking, pelagic in habit.

Similar caution must be taken in dealing with the Copepoda, an order of Crustacea that is essentially pelagic in habit. The only species that has been regarded as undoubtedly abysmal is *Pontostratiotes abyssicola*, a form whose carapace and antennæ are armed with exceedingly long and strongly toothed spines, and was found in the mud brought up by the trawl from a depth of 2,200 fathoms.

Calamus princeps, the largest species of its genus of a deep reddish brown colour, may also belong to the fauna of the deep sea, but we have less evidence concerning the habitat of *Hemicalanus aculeatus*, *Phyllopus bidentatus*, and some of the Euchætæ.

The Amphipoda seem to be but poorly represented in the fauna of the abyss; in fact it may be considered to be still an open question whether any Amphipods habitually live in very deep water.

In the reports on the 'Challenger' Amphipoda, the Rev. T. R. Stebbing states that thirty-one specimens are known to come from great depths, but it would be more correct to say that these specimens were found in the dredges and trawls that had been lowered into the great depths. It should be noticed, however, that some of these specimens do show characters that suggest, at any rate, that they come from deep water. Thus the genus *Lanceola*, for example, is characterised by the smallness of the eyes and a soft membranous integument, while *Cystisoma spinosum*, found in a dredge that had been at work at a depth of over a thousand fathoms, has very large eyes.

In his report on the Crustacea of the 'Norske Nordhavns' expedition, Professor Sars gives a full description of many species of Amphipoda brought by the dredge from depths of over 1,000 fathoms, and nearly all of these were found to be quite blind.

The form that seems to be most peculiar to the great depths of the Northern Ocean is *Harpinia abyssi*. It was found at no less than fifteen different stations at depths ranging from 350 to 2,215 fathoms, and is characterised by its large size and the total absence of eyes.

Another point that should be considered in coming to any conclusion on the supposed habitat of such forms, is the similarity or dissimilarity of widely distributed species.

I have had occasion to point out in a previous chapter the general similarity of the abysmal fauna all over the world, a very striking phenomenon, commented on by almost every naturalist who has had a wide experience of this kind of investigation.

Among the Amphipoda we have a very striking example of this. The species *Orchomene musculosus* was taken by the 'Challenger' off the southern part of Japan at a depth of 2,425 fathoms, the bottom being red clay and the temperature 35·5° Fahr. The species *Orchomene abyssorum* was taken off the east coast of Buenos Ayres at a depth of 1,900 fathoms, the bottom being blue mud and the temperature 33·1° Fahr. To the description of this last-named species Mr. Stebbing adds, 'had this species been taken within reesonable distance of *O. musculosus*, the resemblance is so great that one might have been tempted to disregard the points of difference as due to some other cause than difference of species.'

Such a striking similarity between two species living so far apart from one another may, when we

THE ARTHROPODA OF THE DEEP SEA

take into consideration the depth, the character of the bottom, and the temperature from which they are supposed to have been dredged, be taken to support very strongly the view that these species are really abysmal in habit.

Among the Isopoda we have several very characteristic forms—no fewer than nine distinct genera peculiar to the abysmal zone have been described by Beddard—and of these two, *Bathynomus* and *Anuropus*, are to be regarded as types of sub-families. They seem to be very unevenly distributed over the floor of the ocean, some regions, such as the whole of the Central and Southern Atlantic and the Central and Western Pacific, produce none; whilst the waters of the east coast of New Zealand, the Crozets, and others, produce a great many varieties. Many of the deep-sea Isopoda exhibit characters that are usually associated with the bathybial life. Thus, according to Beddard, thirty-four of the deep-sea species are totally blind, and eighteen have well-developed eyes. In four species there are eyes which are evidently degenerating. If we compare, for instance, the structure of the eye of *Serolis schythei*, a species found in shallow water ranging from 4 to 70 fathoms, with the eyes of *Serolis bromleyena*, a species living in deep water ranging from 400 to 1,975 fathoms, we cannot fail to see that the latter are undergoing a process of degeneration; the retinulæ and pigment being absent, and nothing left of the complicated structure of the Isopod eye but the remnants of the crystalline cones and corneal facets (see figs. 4 and 5, p. 74).

Taking the genus *Serolis* alone, it has been said ' that in all the shallow-water forms the eye is relatively small but very conspicuous from the abundant deposition of pigment; in all the deep-sea forms, with the exception of *S. gracilis*, where the eye seems to be disappearing, it is relatively larger but not so conspicuous, owing to the fact that little or no pigment is present.'

In many groups of animals it has been shown that some of the deep-sea species are relatively much larger in size than the shallow-water species, and that others, more rarely, are much smaller, the abysmal fauna reminding us in this respect of the characters of the alpine flora.

The Isopoda show many examples of this largeness in size, thus *Bathynomus giganteus*, dredged by Professor Agassiz off the Tortugas at a depth of over 900 fathoms, reaches the enormous size, for an Isopod, of 9 inches (fig. 15). *Stenetrium haswelli*, again, is larger than any of the shallow-water species of the genus, and the same remark applies to the deep-sea species of the genus *Ichnosoma*, while *Iolanthe acanthonotus*, from a depth of nearly 2,000 fathoms, is considerably larger than most of the shallow-water Asellidæ.

There is another very common character of deep-sea Crustacea that is also well exemplified in the group of the Isopods, and that is the extraordinary length and number of the spines covering the body.

I have already referred to this character in the supposed deep-sea Copepod *Pontostratiotes abyssicola*, and I shall have again to refer to it in treating of the Decapoda and other groups of the Crustacea.

Besides its enormous size *Bathynomus* possesses some other characters that may be correlated with its deep-sea environment. The respiratory organs are quite different from those of other Isopods; instead of being borne by the abdominal appendages, they are in the form of branched outgrowths from the body-wall containing numerous blood-lacunæ, and the appendages simply act as opercula to cover and protect them. The eyes of the Bathynomus too are remarkably well developed, each one bearing 4,000 facets, and they are directed not dorsally as in the Cymothoadæ, but ventrally. The cause of these curious modifications of structure in Bathynomus is by no means clear, but it is quite probable that they are connected with the conditions of pressure and light in the deep sea. It is a remarkable fact that the other deep-sea Isopods do not exhibit precisely these modifications, and it might be supposed that the same causes would produce the same or similar effects on the structure of animals belonging to the same order. That is perfectly true, but we cannot yet determine how long ago any one species has taken to a deep-sea life, or what length of time, in other words, these conditions have been at work in modifying the structure of the organism. A recent immigrant into the abyss will naturally exhibit closer affinities with its shallow-water allies than those that have dwelt in the region since secondary or tertiary times. If we take this into consideration we should expect to find considerable differences occurring between deep-sea species of the same order, which is precisely what we do find.

Concerning the Cirripedia, that curious group of profoundly modified Crustacea that includes the

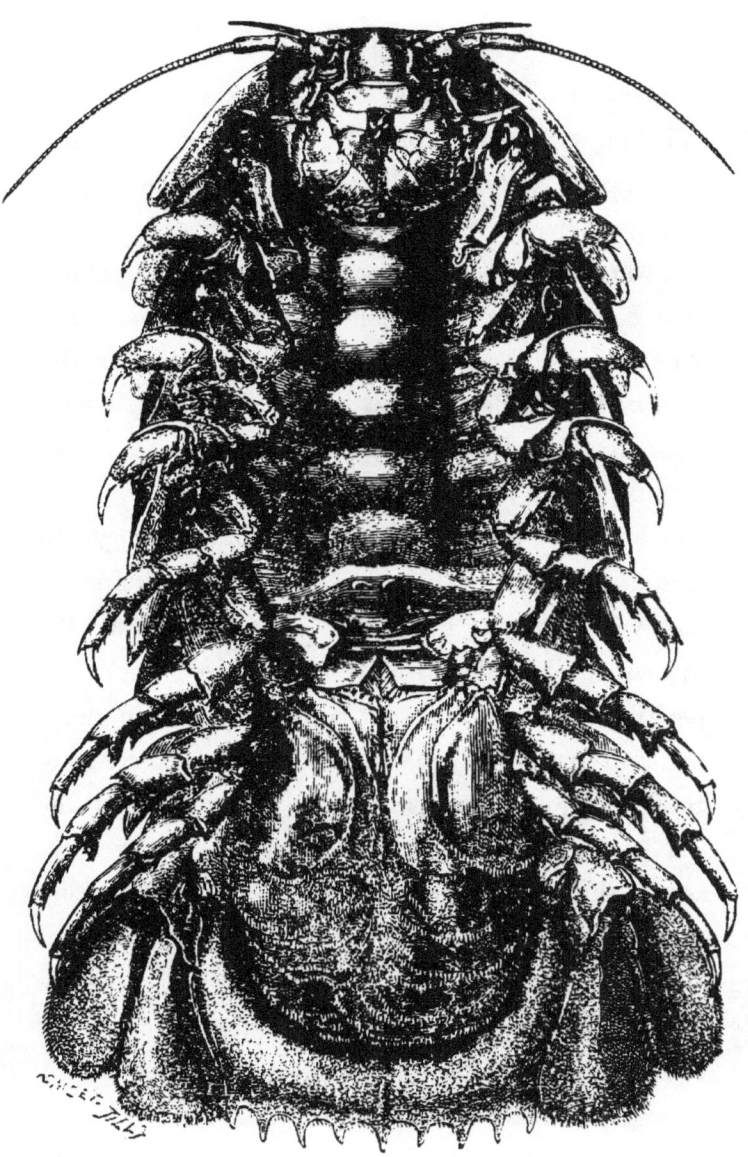

FIG. 15.—*Bathynomus giganteus.* From a depth of 1,710 metres. (From Filhol.)

barnacles and acorn shells, Dr. Hoek writes in the 'Challenger' monograph:—

'Though unquestionably by far the greater part of the known Cirripedia are shallow-water species, and though some of the species are capable of living at a considerable variety of depths, as, for instance, *Scalpellum stroemii*, yet it must be granted that the number of true deep-sea species of Cirripedia is very considerable.' Only two genera, however, occur in depths of over 1,000 fathoms, and these—*Scalpellum* and *Verruca*—occur also as fossils in secondary and tertiary deposits. The oldest of all fossil cirripedes, however, namely, *Pollicipes*, never occurs, at the present day, in deep water, but is purely littoral or neritic in habit. But what is perhaps more interesting still is the fact, that, when we come to compare the living and the fossil species, we find that in the one genus (*Scalpellum*) the deep-sea forms have preserved the more archaic characters, and in the other (*Pollicipes*) the shallow-water forms.

Here then we are presented with a veritable puzzle for which we can at present frame no manner of answer. Pollicipes on the one hand—like Lingula among the brachiopods—has been able to maintain itself almost unchanged amid the tremendous struggle for life of the shallow water of the tropics ever since the Lower Oolite epoch; while Scalpellum, on the other hand, has either become profoundly modified, or been driven into the abysmal depths of the ocean.

The group of the Thoracostraca, or stalk-eyed Crustacea, including lobsters, crabs, hermit crabs, prawns, and shrimps, is well represented in the deep sea. Most of them are characterised by being quite blind (in many cases even the eye-stalks are obliterated), by being protected with a dense covering of spines, by the thinness of their shells, and by their bright red or carmine colour.

The order Stomatopoda is almost entirely confined to the shallow waters of the tropical or temperate shores. Not a single species is known to inhabit the deep sea, and only a very few specimens have been captured in more than a few fathoms of water.

The Schizopoda, however, present us with many curious abysmal forms. Most of the genera of this order belong to the pelagic plankton, and many of them are known to possess the power of emitting a very strong phosphorescent light. Several genera, however, such as *Gnathophausia*, *Chlaraspis*, *Eucopia*, *Bentheuphausia*, &c., never seem to leave the great depths of the ocean, and nearly all of these genera are distinguished by being quite blind or possessing very much reduced or rudimentary eyes.

If we compare, for example, the pelagic *Euphausia latifrons* (fig. 16) with the nearly allied but abysmal

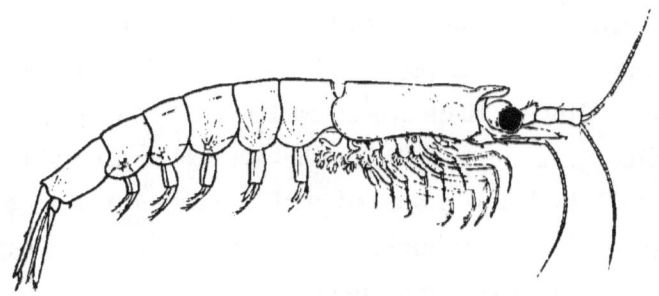

FIG. 16.—*Euphausia latifrons*, from the surface of the sea. (After Sars.)

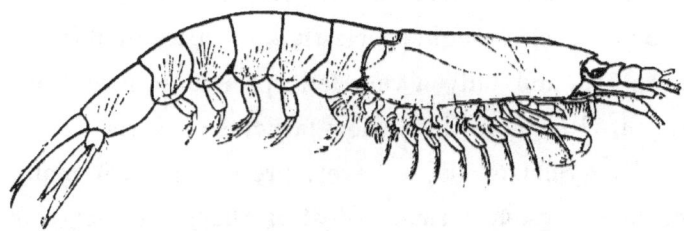

FIG 17.—*Bentheuphausia amblyops*, from 1,000 fathoms. After Sars.

Bentheuphausia amblyops (fig. 17), the difference in this respect between a Schizopod living in the sunlight and one living in the darkness of the deep-sea is very apparent.

The pelagic Schizopoda are usually quite pale and transparent; the deep-sea forms on the other hand are frequently if not invariably of a bright red colour, as is the case with many other deep-sea Crustacea to which reference will be made later on.

Passing on to the group of the Decapoda, we find that the most interesting of all the abysmal crayfish is the family of the Eryonidæ; indeed, in some respects the discovery of these curious forms may be reckoned among the most valuable results of the 'Challenger' Expedition. They are characterised by the dorsal depression of the anterior part of the cephalothorax, the absence of a rostrum, and the absence or very rudimentary condition of the eyes (fig. 18).

Their nearest relations seem to be certain genera of Crustacea that are found in jurassic strata, in the lias, and more particularly in the lithographic slates of Solenhofen.

They have a very wide bathymetrical range extending from a depth of 250 fathoms (*Polycheles crucifera*) to a depth of 2,000 fathoms (*Willemoesia*).

But there are many other curious forms of the macrurous crustacea that deserve a passing mention. The graceful *Nematocarcinus gracilipes*, distinguished by the extraordinary length of the antennæ and last four pairs of legs, these appendages being

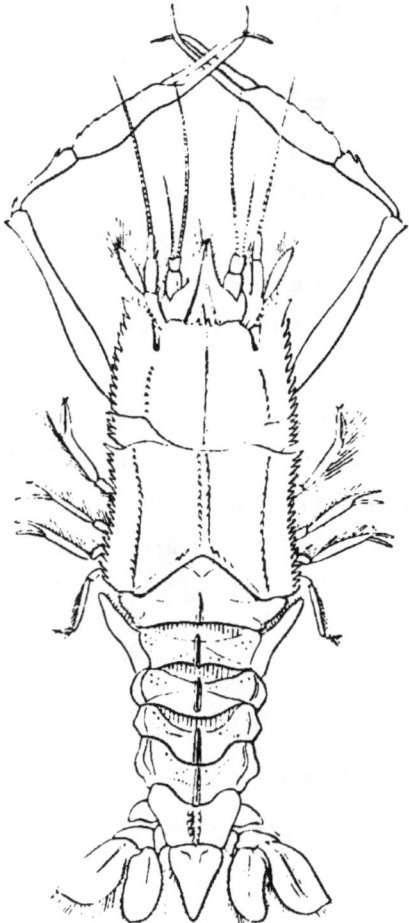

FIG. 18.—*Polycheles baccata*, one of the Eryonidæ. The eyes and eye-stalks are absent, and the margin and sides of the carapace armed with spines. (After Spence Bate.)

three or four times the length of the body. is by no means rarely met with in depths of over 400 fathoms.

The genus *Glyphus* captured by the 'Talisman' is remarkable for the development of a peculiar pouch-like arrangement on the abdomen for the protection of the larvæ during the younger stages of their existence.

The proof of the existence of a peculiar cray-fish, *Thaumastocheles zaleuca*, at a depth of 450 fathoms, was one of the most important contributions to carcinology made by the 'Challenger' Expedition. The chelæ of this remarkable form are of great but unequal length and armed with long tooth-like spines giving it an appearance not unlike that of the jaws of some carnivorous fish. The shell is soft and the abdomen broad and flattened. There are no eyes nor even eye-stalks, but ' in front of the carapace,' as Sir Wyville Thomson remarks, ' between the anterior and upper edge and the insertions of the antennæ, in the position of the eyes in such forms as *Astacus fluviatilis*, there are two round vacant spaces, which look as if the eye-stalks and eyes had been carefully extirpated and the space they occupied closed with a chitinous membrane.' The deep-sea prawn, *Psalidopus*, recently taken in 500 fathoms of water by the

'Investigator,' affords us an example of a common bathybial character, the whole body being covered with an extraordinary array of sharp needle-like spines.

Among the crabs many curious forms have been found in deep water extending down to depths of over 2,500 fathoms. They are nearly all characterised by blindness and a remarkable development of tooth-like spines covering the carapace and limbs.

The remarkable *Lithodes ferox*, from a depth of from 450 to 800 fathoms, is perhaps the most perfectly armed crab—in the way of spines—that exists. Every part of the body and limbs is so covered with spines that one has to be extremely careful in handling even a dead specimen.

This is only one of the many examples that might be given to illustrate this curious feature of the deep-sea Crustacea. Among the crabs alone we have such forms as *Galathodes Antonii*, *Pachygaster formosus*, *Dicranodromia mahyeuxii* covered with a fierce armature of spines or bristles; but there are nevertheless some species in which this character is not particularly noticeable, and in these we usually find some other protection against their enemies. An interesting example of this has been described by A. Agassiz in a crab allied to the Maiadæ, 'in which the dorsal face appears like a bit of muddy area covered by corals, with a huge white arm resembling a fragment of an Isis-like gorgonian.' It is evident that this is a case in which the animal is protected by its resemblance to the surroundings.

The hermit crabs of the abyss, too, are not usually characterised by any very great development of spines. They find their protection in the shells they inhabit. Some of the deep-sea hermit crabs carry about with them on their shells a sea anemone, as we find to be frequently the case among the shallow-water species. *Pagurus abyssorum*, from a depth of 3,000 fathoms, is an example of this.

In cases where there is a scarcity of gasteropod shells the hermit crabs are obliged to find some other form of protection for their bodies. The 'Blake' found in the West Indies a hermit crab that had formed for itself a case of tightly compressed sand, and another curious form, named *Xylopagurus rectus*, makes its home in pieces of bamboo or in the holes in lumps of water-logged wood.

The last group of the Arthropoda we need refer to is that of the Pycnogonida, those curious creatures seemingly made up entirely of legs, and by some naturalists considered to be related to the Crustacea and by others to the scorpions and spiders.

Like the Brachiopoda the Pycnogonida are not usually found in greater depths than 500 fathoms. Out of the twenty-seven known genera, only five extend into the abyss, and not one of these can be called a true deep-sea genus.

There are three genera, *Nymphon*, *Collosendeis*, and *Phoxichilidium*, that show a very wide distribution over the floor of the ocean, and are capable of existing at the greatest depths, and of these the species of the genus Nymphon have a truly remarkable range extending from the shore to a depth of 2,225 fathoms.

'As a rule,' says Hoek, ' the deep-sea species are slender, the legs very long and brittle, and the surface of the body smooth.' They have further, either no eyes at all or rudimentary eyes without pigment, and in many cases—as, for example, *Collosendeis*—they are distinguished for reaching to a gigantic size compared with their shallow-water relatives.

The Tunicata is the group of animals that includes all those curious vegetable-like organisms found upon our coasts that are familiarly known as sea-squirts, or Ascidians, besides the salps, pyrosomas,

FIG. 13.—*Colossendeis arcuatus*, from a depth of 1,500 metres. (After Filhol.)

and the microscopic appendicularias of the pelagic plankton.

Notwithstanding the apparent simplicity of their adult structure, naturalists are now agreed that they must be removed from the Mollusca, with which they have hitherto been most frequently associated, and placed in the group of the Vertebrata. It is the study of embryology that has led to this unexpected conclusion, for we find, when we study the larval forms, that they possess both a notochord and gill-slits, two features that are characteristic of the group of the Vertebrata.

The species of the group Perennichordata, which includes all those Tunicates that possess a notochord persistent through life, are chiefly pelagic in habit, the little creatures, rarely more than two or three millimetres in length, swimming or drifting about with the sagittas, copepods, ctenophores, and medusæ that compose the pelagic plankton. Fol has recently described a gigantic form belonging to this group, reaching a size of thirty millimetres in length, called *Megalocercus abyssorum*, which he dredged from a depth of 492 fathoms; and other species have been recorded down to a depth of 710 fathoms in the Mediterranean Sea.

Among the simple Ascidians we find no family that is peculiar to deep water; but the Cynthiidæ and Ascidiidæ both contain genera that are abysmal, and the Molgulidæ have one species, *Molgula pyriformis*, that extends into the abysmal zone to a depth of 600 fathoms.

In the genus *Culeolus* and in *Fungulus cinereus* and *Bathyoncus*, all deep-water Ascidians, there is a very curious modification of the branchial sac, the stigmata being apparently not formed, in consequence of the suppression of the fine interstigmatic vessels. This peculiar feature is only found in the deep-sea simple Ascidians and, as we shall see presently, in one species of the deep sea compound Ascidians, but it is not apparently an essential character of those living in the abysmal zone, notwithstanding the fact that it is found in such widely separated genera; for *Corynascidia*, *Abyssascidia*, and *Hypobythius*, living in depths lying between 2,000 and 3,000 fathoms below the surface, have branchial sacs of the ordinary type. Professor Herdman is of opinion that this simple form of branchial sac is not a primitive form, but most probably a modification of a more complicated type.

In *Culeolus Murrayi* there is a remarkably abundant supply of blood-vessels to the tunic, and these send special branches to a number of small papilliform processes on its outer surface. This system of highly vascular processes probably constitutes, as Professor Herdman suggests, an additional or complementary respiratory apparatus. All these modifications of the branchial system are of particular interest, for we find so many instances of a similar kind among the inhabitants of very deep water. I need only refer here to the modifications of this system in the Isopod *Bathynomus* already referred to (p. 129), and to the reduction in the number of the gills of many of the deep-sea fishes. Why there should be such modifications is a question upon which the physical and natural history investigations of the conditions of life in the great depths of the ocean at present throw no light.

In a previous chapter I have referred to the fact that many of the bathybial animals are characterised by being stalked. Among the simple Ascidia we find many examples of stalked kinds living in deep water, such as *Culeolus* and *Fungulus*, but also several exceptions, such as *Bathyoncus*, *Styela bythii*, and *Abyssascidia*, that are sessile. It is a noteworthy fact, however, that the genus that has the most deep-sea species—namely, *Culeolus*—is a genus that is provided

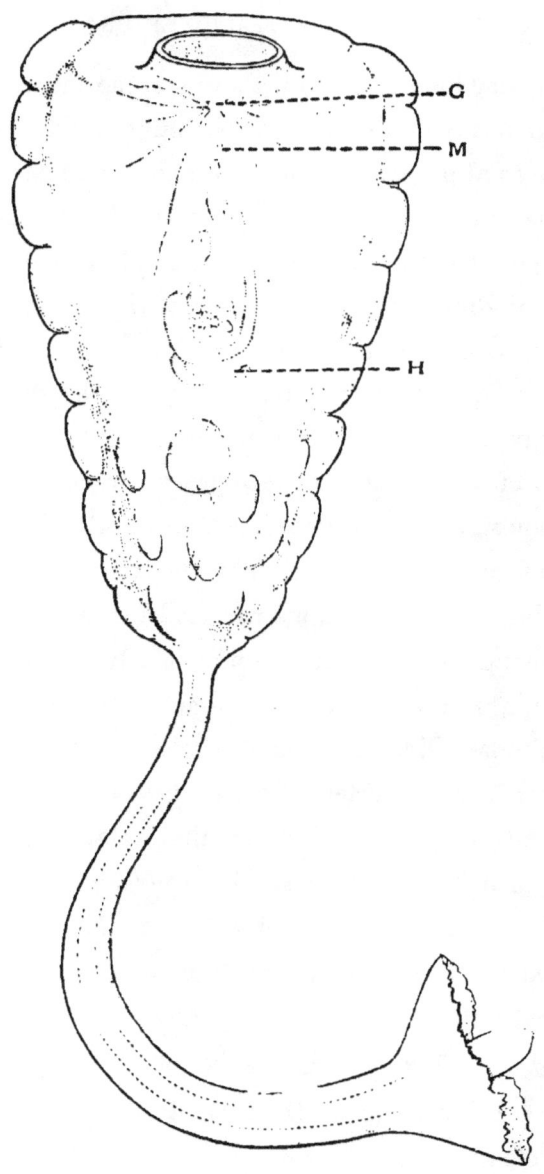

FIG. 20.—*Hypobythius calycodes.* G, nerve ganglion; H, heart; M, the position of the atriopore. The large opening on the upper side is the mouth. (From a drawing by Professor Moseley in Herdman's 'Tunicata of the "Challenger" Expedition.')

L

with a very long stalk. Furthermore, the only known stalked forms of the very large family Ascidiidæ are the abysmal genera *Corynascidia* and *Hypobythius*.

The most remarkable character of the genus *Hypobythius* is the simple condition of its branchial sac, reminding one of the structure of this organ in the shallow-water genus *Clavelina*. 'There are no folds and there are no internal bars,' to quote the description given by Professor Herdman; 'only a single system of vessels can be recognised, branching and anastomosing so as to form a close network, the small rounded meshes of which are the stigmata. The tentacles and dorsal lamina cannot be made out.'

Among the compound Ascidians only four families extend into the abysmal zone, namely, the Botryllidæ, Polyclinidæ, Didemnidæ, and Cœlocormidæ, and of these only one species, *Pharyngodictyon mirabile*, of the family Polyclinidæ, extends into water of greater depth than 1,000 fathoms. In *Pharyngodictyon* we find the same curious simplification of the branchial sac that we have just referred to in the genera of simple Ascidians, *Culeolus*, *Fungulus*, and *Bathyoncus*. *Cœlocormus Huxleyi* from a depth of 600 fathoms is a very peculiar form and the type of a separate family, the Cœlocormidæ.

The free-swimming Tunicata included in the group *Ascidiæ salpiformes*, which contains the genus *Pyrosoma*, and the order Thaliacea containing the salps, are in all probability mainly confined to the surface waters. A few specimens of *Pyrosoma* were captured by the 'Challenger' dredges which came up from very deep water, but it is doubtful at what point in the journey to the surface the specimens entered the net.

The most remarkable form of free-swimming Tunicate that has come to light is *Octacnemus bythius*, a form that is probably allied to *Salpa*. It was found twice, once in the dredge that came from a depth of 1,070 fathoms, and once from 2,160 fathoms. The tunic of the animal is gelatinous and hyaline, but the most important feature it possesses is an imperforate membrane separating the branchial sac from the peribranchial cavity. Octacnemus, in other words, possesses no true stigmata, these structures being represented only by little pits in the walls of the branchial sac. This curious and extremely interesting modification of the respiratory organs points very strongly to the conclusion that Octacnemus is truly a deep-sea animal.

CHAPTER VIII

THE FISH OF THE DEEP SEA

OF all the groups of animals that constitute the deep-sea fauna, the fish show the greatest number of peculiarly abysmal characters. Being much more highly differentiated than the invertebrates, they possess more organs liable to undergo modifications of colour, size, and structure, and consequently we are able to point to a great many more features characteristic of deep-sea fish than we can do in any other group of animals.

The first point that calls for remark in the consideration of the fish fauna of the deep sea is the almost entire absence of ancient and primitive types. The Elasmobranchii, including the Sharks, Rays, and Chimæra, constituting the order that from anatomical embryological grounds is always regarded by naturalists as the most primitive order of this class, is represented in very deep water by only one species.

Raia hyperborea and *Chimæra monstrosa*, it is true, just enter into the abysmal zone. but *Chimæra affinis* is the only Elasmobranch that extends to depths of over 1,000 fathoms.

The Ganoidei too, the order that in palæozoic and mesozoic times was so rich in genera and species, is entirely absent from the abysmal zone, not a single representative having been found at any time by any of the deep-sea expeditions.

The Dipnoi, that remarkable order including the three fresh-water genera, *Ceratodus* from Australia, *Lepidosiren* from Brazil, and *Protopterus* from West Africa, has no representative and no ally in the deep waters of the ocean.

The fishes of the deep sea, in fact, with only one or two exceptions, all belong to the Order Teleostei, the most modern and most highly differentiated order of the class, the families that are most fully represented being the Macruridæ and then the Ophidiidæ and Gadidæ, and the Berycidæ.

At the limits of the katantic and abysmal zones, a large number of families of Teleosteans entirely disappear, and as we approach the deepest parts of the ocean, the number of fish that are found is considerably reduced. As Dr. Günther very wisely remarks, this diminution in the number may be due to the difficulty of capturing fishes at great depths, a difficulty which increases in proportion to the depths at which the dredge is worked. But it must also be regarded as evidence of the actually diminished variety of fishes.'

It may be interesting to the reader to give Dr. Günther's table of the number of species found at different depths, as it shows, among other things, the marked change that occurs in the character of the fauna in passing from the katantic to the abysmal zone.

Between 100–300 fathoms,	232 species
,, 300–500 ,,	142 ,,
,, 500–700 ,,	76 ,,
,, 700–1,500 ,,	56 ,,
,, 1,500–2,000 ,,	21 ,,
,, 2,000–2,900 ,,	23 ,,

As regards the general character presented by the deep-sea fishes, I have already pointed out in the chapter dealing with the general characters of the deep-sea fauna, the peculiarities in the size of the eyes, the colours and markings of the body, and the texture of the bones and muscles. There are, however, a few more characters of which mention must be made.

Notwithstanding the fact that all the abysmal fibres are carnivorous and must consequently be

THE FISH OF THE DEEP SEA 151

capable—in the great number of cases—of rapid and vigorous movement, the muscles of the trunk and tail are usually thin, and the fascicles loosely connected with one another.

Deep-sea fish are not characterised by an absence of the swimming bladder. This organ occurs just as frequently and in the same families as in the shallow-water fauna, but we do not know whether it possesses any special peculiarities or not, as it is usually so ruptured and destroyed by the change of pressure it undergoes in being brought to the surface, that it is impossible to make any thoroughly accurate investigation of its anatomy and relations.

The extraordinary development of glands in the skin which secrete mucus, and the presence in many forms of very highly specialised organs for emitting phosphorescent light, are characters of the deep-sea fish fauna, to which I have referred in a previous chapter.

As with the Tunicates, some of the Crustacea and other groups, the fish of the abysmal zone show curious modifications of the respiratory system. The gill laminæ of these animals are not only reduced in number, but appear to be short and shrunken. It is possible, of course, that during life they may end in fine delicate points which are broken off or ruptured during their capture, but still the horny rods that support them are shorter than they are in shallow-water forms, and the general evidence of their structure tends to show that they have undergone profound modifications in the change to the conditions of deep-sea life.

An extremely common and almost general character of deep-sea fishes is the black coloration of some of the body cavities; this is limited to the pharynx in many of the fishes that live about the hundred fathoms limit, but the colour is more intense and spread all over the oral, branchial, and peritoneal cavities in typical deep-sea forms. It may seem very difficult at first to account for this remarkable development of black pigment in parts of the body that are not usually, and, in some cases, cannot at any time be exposed to view. It is obvious that it cannot be functional as a hiding colour, either in offence or defence. But it is quite possible that it is due to some modification of the function of excretion. It is well known that in many cases of disease or injury to the kidneys in vertebrates, the colour of the skin is affected, and every one recognises now the fact that in many invertebrates the colour of the skin is greatly dependent upon the function of the secretion of the urates.

It would at least be interesting to know if this dark coloration of the mucous membranes is in any way correlated with any modification of the structure or function of the kidneys. At present we have no recorded observations on this point, but it is to be hoped that, when we have a sufficient number of specimens brought home from the deep water, a systematic investigation of this subject will be made.

Lastly, it should be pointed out that our knowledge of the abysmal fauna has not, at present, brought to light any evidence that the fish are of an extraordinarily large size. In many groups of animals, as I have frequently pointed out in the last few chapters, the large and gigantic species or specimens are only found in the abyss. This may also be the case with fishes, but we have no evidence that it is so. The only methods that have been used at present for the investigation of the fauna living on or near the floor of the deep oceans, are not of a kind to lead to the capture of really large fish. That they may exist is highly probable, but all that we know at present is, that the fish with which we are acquainted living at great depths are not in any way remarkable for their great size.

Of the only two Elasmobranchs, one, namely *Raia*

hyperborea, has been found in water extending from 400 to 608 fathoms in depth. Only four specimens have yet been taken, one by the Norwegian expedition off Spitzbergen and three by the 'Knight Errant' off the northern coasts of Scotland. It is interesting to find that this, the only deep-sea species of the Rays, shows some striking peculiarities. 'The teeth are remarkably slender,' says Günther, 'small, irregularly and widely set, different from those of other British Rays. In young specimens at any rate those of the male do not differ from those of the female. The mucous membrane behind the upper jaw forms a pad with a lobulated surface. The mucous cavities of the head are extremely wide, and finally the accessory copulatory organs have a spongy appearance, and are flexible, the cartilage by which they are supported being a simple slender rod.'

The other Elasmobranch, that extends into very deep water, is *Chimæra affinis*, a species which can hardly be distinguished from the better known *Chimæra monstrosa*, a fish that itself very frequently wanders within the limits of the abysmal zone.

Among the Teleostei, the family Berycidæ has several representatives in the deep water. They are small fish rarely exceeding four inches in length, with large heavy heads, with functional but small eyes, and an abundant supply of large mucous glands on the skin.

Melamphaes beanii, belonging to this family, has been captured at the enormous depth of 2,949 fathoms.

Bathydraco antarcticus, belonging to the family Trachinidæ, from a depth of 1,260 fathoms, is an example of a true abysmal fish possessing very large eyes.

The Pediculati, the family of the anglers, is represented at depths of over 2,000 fathoms by the interesting form *Melanocetus Murrayi*. The eyes are very small indeed, the mouth huge and armed with long uneven rasp-like teeth. At the end of the fishing-rod tentacle hanging over the mouth, there is an organ that has been supposed to be capable of emitting a phosphorescent light. This curious modification of the red worm-like bait of the common shallow-water angler into a will-o-the-wisp lantern attracting little fishes to their destruction in the deadly jaws of the *Melanocetus* is one of the most interesting adaptations that has been brought to light by our study of the deep-sea fauna.

Several species of the family Lycodidæ occur in

the abysmal zone, but they do not possess any features that call for special mention in this place.

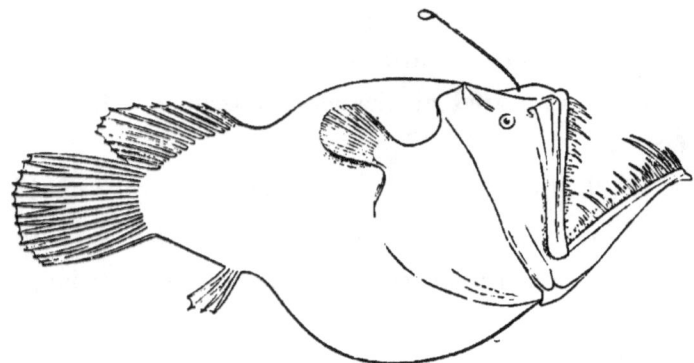

FIG. 21.—*Melanocetus Murrayi*, 1,850–2,450 fathoms. (After Günther.)

The family Ophidiidæ contributes very largely to the fish fauna of the abyss. Some of the deep-water genera, such as *Neobythites*, have a wide bathymetrical distribution extending from 100 fathoms to depths of over 2,000 fathoms, but others, such as *Bathyonus*, *Typhlonus*, and *Aphyonus*, only occur in depths of over 1,000 fathoms.

The body is usually elongate and slender, ending in a pointed tail, the head large and heavy, and the eyes, in the genera confined to the abysmal zone, usually so far degenerated that they are not visible at all from the outside.

The Macruridæ form a family that contributes

very largely to the deep-sea fauna; no fewer than twenty-six different species are known to occur within the limits of the abysmal zone. Not only do the Macruridæ contribute a large number of different species, but they probably occur, in some districts at any rate, in vast numbers.

During the voyage of the 'Talisman,' for example, the French naturalists caught in one haul of the dredge off the coast of Morocco in 500 fathoms of water no fewer than 134 fish, of which number 95 belonged to the family Macruridæ.

They are usually small fish, measuring from a few inches to two feet in length, with a body terminating in a long compressed tapering tail and covered with spiny, keeled, or striated scales.

The Pleuronectidæ or flat fish are not, as a rule, found in the abysmal zone; one species, however, *Pleuronectes cynoglossus*, was found by the American ship 'Blake' to extend into 732 fathoms of water.

The families Sternoptychidæ and Scopelidæ are of particular interest to us, as almost all the genera they contain belong either to the pelagic or abysmal zones, and lend support to the view enunciated by Moseley, that the deep-sea fauna has, partly at any rate, been derived from the fauna of the pelagic zone.

They are nearly all small slender fish with delicate and frequently semi-transparent bodies, large gaping mouths armed with numerous long irregular teeth, and frequently provided upon the head and sides of the trunk with rows of eye-like phosphorescent organs.

These families, and others that have still to be referred to, belong to the group of Teleostei that is called Physostomi, the name referring to the open communication that usually exists in all these families between the swimming bladder and the alimentary canal. It is a remarkable fact that in none of the deep-sea representatives has this open communication been discovered. It is true that many specimens are, when examined, so lacerated by the diminution in pressure as to render anatomical study a matter of difficulty, but still a fair number of uninjured well-preserved specimens have now been examined and the duct has not been found.

Of the family Sternoptychidæ, *Gonostoma microdon* has a most remarkable distribution. It has been found at numerous stations in both the Pacific and Atlantic Oceans at depths ranging from 500 to nearly 3,000 fathoms of water.

The Scopelidæ are represented by some very

extraordinary types. The genus *Bathypterois*, for example, occurring in depths ranging from 500 to 2,500 fathoms, is characterised by the development of enormously long pectoral fins to serve probably as organs of touch. 'The rays of the pectoral fin,' says Dr. Günther, 'are much elongated. The ventral fins abdominal, with the outer rays prolonged, eight-rayed. . . . Gill rakers long.' They are further characterised by the absence of any true phosphorescent organs and the smallness of their eyes.

There can be little doubt, I think, that in these fishes the sense of touch or taste to a great extent takes the place of the sense of sight in other Scopelids. Not being provided with well-developed eyes or phosphorescent organs to attract their prey, the pectoral fins and the outer rays of the pelvic fins have become elongated and provided with special sense organs for searching for their food in the fine mud of the floor of the ocean.

These long pectoral rays must have a very curious appearance in the living fish. Mr. Murray observes: ' When taken from the trawl they were always dead, and the long pectoral rays were erected like an arch over the head, requiring considerable pressure to make them lie along the side of the body; when erected they resembled the Pennatulids like Umbellula.' Filhol considers that when the fish is progressing through the obscurity of the abyss it probably carries these organs directed forward, seeking with them in the mud for any worms or other animals upon which it preys, or receiving through them warning of the approach of an enemy from whom it is necessary to make an immediate escape. One of the most remarkable of the deep-sea fish is closely related to Bathypterois, namely *Ipnops Murrayi*, living in depths of over 1,000 fathoms. It is about five inches long, of a yellowish brown colour, with an elongated subcylindrical body covered with large thin deciduous scales. There are no phosphorescent organs of the ordinary type met with in the Scopelidæ, but upon the upper surface of the head there is found a pair of organs somewhat resembling the ordinary eyes of fishes but devoid of retina and optic nerve, that, from the researches of Moseley, seem to be undoubtedly organs for emitting light. 'The organs are paired expanses, completely symmetrical in outline, placed on either side of the median line of the upper flattened surface of the head of the fish, extending from a line a little posterior to the nasal capsules nearly to a point above the posterior extremity of the cranial cavity.' They are covered by the upper walls of the skull, which is extremely thin and completely transparent in the region lying over them. 'They are membranous structures 0·4 mm. in thickness marked by hexagonal areas about 0·04 mm. in diameter. When their surface is viewed by reflected light the appearance is that of a number of glistening white isolated short columns standing up in relief from its basal membrane.' Each hexagonal column is composed of a number of transparent rods disposed side by side at right angles to the outer surface of the organ, with their bases applied against the concave surface of a large hexagonal pigment cell, one of which forms the basis of each hexagonal column. It is still very doubtful what are the true homologies of this extraordinary phosphorescent organ, but Moseley was of opinion that, ' on the whole, it seems not unlikely that the remarkable head organs of Ipnops may be regarded as highly specialised and enormously enlarged representatives of the phosphorescent organs on the heads of such allied Scopelidæ as *Scopelus rafinesquii* and *Scopelus metopoclampus*. It may be conceived that in *Ipnops* the supra-nasal and sub-ocular phosphorescent organs of these species on either side have united and become one with the result of the total obliteration of the eye.'

Most of the species of the genus *Scopelus* are undoubtedly pelagic in habit, descending during the day to depths of semi-darkness but rising at night to the surface waters. It is not certain how many of the known species occasionally or habitually dwell in very deep water, but there seems to be no doubt that two species at least—*S. macrolepidotus* and *S. glacialis*—belong to the abysmal zone. Both of these species were found in dredges that had been at work in depths of over 1,000 fathoms and showed signs when examined of having been brought from the abyss.

The Stomiatidæ are almost entirely confined to water from 450 to nearly 2,000 fathoms in depth. They may be distinguished from the Scopelidæ by the long hyoid barbel close to the symphysis of the lower jaw, but like many of the genera of that family they have wide gaping mouths armed with a profusion of vicious looking teeth and a series of luminous spots on the sides of the head and body. (*See* Frontispiece.)

In *Eustomias obscurus*, found in depths of over 1,000 fathoms in the Atlantic by the 'Talisman,' the barbel is provided with a terminal swelling, shaped like a dumb-bell, which may be capable of emitting a phosphorescent light and serve the animal as a lure for the attraction of its prey. The genus *Malacosteus*, too, presents us with some of the most remarkable forms that are found in the abysmal zone. The mouth is of enormous size and the integuments of the abdomen present very definite longitudinal folds, leaving no doubt that this fish is able, like several others living in deep water, to swallow prey of an enormous size.

But a perfectly unique structure in this fish 'is a thin cylindrical muscular band which connects the back part of the mandibular symphysis with the hyoid bone. It is probably the homologue of a muscular band which, in other Stomiatids, stretches on each side from the mandible to the side of the hyoid, the two bands coalescing into an unpaired one in *Malacosteus*. It is, in the present state of preservation, much elongated, like a barbel, but during life it is notably contractile, and serves to give to the extremity of the mandible the requisite power of resistance when the fish has seized its prey, as without such a contrivance so long and slender a bone would yield to the force of its struggling victim.'

Belonging to the family of the Salmons we find one genus *Bathylagus* that is undoubtedly an abysmal form. Although there may be some doubt as to the exact depth at which the specimens were captured, the thinness of the bones, the enormous size of the eyes, and other bathybial characters prove that they must live in very deep water. Closely allied to the salmon and the herrings is the family of the Alepoce-

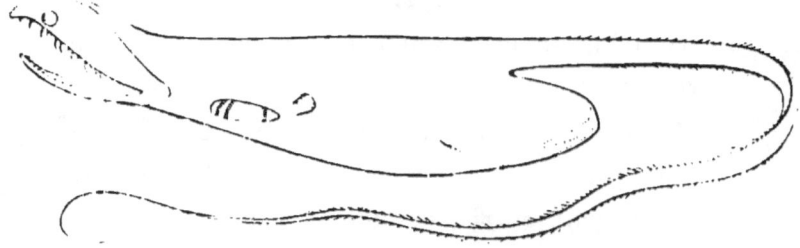

FIG. 22.—*Saccopharynx ampullaceus*; a deep-sea eel, with the head of a large fish, which it has swallowed, showing through the thin integuments of the body. (From Günther.)

phalidæ, a family that contributes several forms to the fauna of the deep sea, but they do not possess any characters that call for special comment. Their vertical distribution varies between 345 and 2,150 fathoms.

The family Halosauridæ contains five species all confined to the abysmal zone. They have long bodies tapering to a finely pointed tail, and the head is provided with a snout that projects considerably in front of the mouth.

Of the family of the Eels there are several representatives in the deep sea. They are characterised by a combination of true eel characters with special modifications due to a bathybial existence. 'To enable them to seize upon prey more powerful than themselves certain organs have undergone a degree of specialisation, as is observed in bathybial members of other families with a similar mode of life; the jaws are exceedingly elongate and the whole gape, the pharynx and stomach capable of enormous distension.' The head is very large, the eye very small and the tail long and tapering (fig. 22).

The lessons we learn from the study of the fishes of the deep sea are particularly instructive. It would take far more space than can be afforded here to fully illustrate all of the points that seem clear to us, but I hope I have said sufficient to show that the fish fauna is made up of genera and species belonging to several widely separate families of the Teleostei; that some of them show, in a very marked way, what may be looked upon as peculiarly bathybial characters, whilst others are but slightly modified from their shallow-water representatives. These facts by themselves lend support to the view that the fauna of the deep sea has been derived from the fauna of shallower water by successive migrations at different periods of the world's history. Those that exhibit in a most marked degree the special bathybial characters are probably those whose immigration took place long ago, whilst those more closely related to shallow-water forms are, comparatively speaking, recent importations. The occurrence of Scopelidæ and Sternoptychidæ in deep water suggests, as Moseley pointed out many years ago, that the fauna is partly derived from the pelagic plankton. But while these points may seem clear to us, there are others that still require much more investigation and consideration. The whole question of the function and use of the phosphorescent organs, the mucous glands, the barbels and elongated fin rays, the mode of life, the deposition of ova and their development, afford problems which in the present state of our knowledge must remain unsolved. Let us hope that in the future there may be a new stimulus given to deep-sea research, and these problems may be again seriously studied and eventually solved.

INDEX

ACTINIARIA, 36, 93
— two remarkable genera of, 15
Aegir, 15
Agassiz, A., on colour of Cœlentera, 65
— on Echinoidea, 101, 103
Agassiz, L., on board the 'Hassler,' 12
'Albatross,' American vessel, 15
Alcyonaria, 95
— phosphorescence of, 81
Amphipoda, 124
Anemones of deep water, 36, 92
— two remarkable genera of, 15
Annelida, 117
Ascidia compositæ, 146
— salpiformes, 147
— simplices, 142
Asteroidea, 104

BARRIERS of temperature, 32
Bathynomus, 129
Beddard, F. E., on Isopoda, 127
Benthos, 53
Berycidæ, 154
'Blake,' American vessel, 12
Blue mud, 42
Brachiopoda, 115
Brachyura, 138
Brisinga, 9, 105
Buchanan's experiment, 19

CARPENTER, P. H., on Crinoidea, 100
Cephalodiscus, 113
Cephalopoda, 120
'Challenger,' H.M.S., voyage of, 12
Cirripedia, 130
Cœlentera, colour of, 65
— of deep sea, 91
Colour of the deep-sea fauna, 59, 66
— of the deep-sea fish, 60
Copepoda, 124
Corals, 94
Crinoidea, 99
Crustacea, 123
— colour of, 63

DARKNESS of the abyss, 22
Diatom ooze, 39
Dipnoi, 149

ECHINODERMA, 99
— colour of, 64
Echinoidea, 101
Eels, 165
Elasipoda, 106
Elasmobranchii, 148, 153
Eryonidæ, 135
Eyes of abysmal animals, 67
— of deep-sea crustacea, 72

168 INDEX

Eyes of deep-sea fish, 69
— of deep-sea mollusca, 71
— of *Genityllis*, 118
— of *Neobythites*, 69
— of *Serolis*, 73

Fenja, 15
'Fish Hawk,' American vessel, 12
Fol and Sarasin's experiments, 25
Foraminifera, 90
Forbes, on the probable existence of a deep-sea fauna, 2
— on zones of distribution, 49

GANOIDEI, 149
Gasteropoda, 119
Gephyrea, 116
Gills of deep-sea fish, 151
Globerigina ooze, discovery of, 5
— distribution and composition of, 37,
Green mud, 42
Gunn, Dr., on the eyes of *Genityllis*, 118
Günther, Dr., on deep-sea fish, 150

HALL, Marshall, 12
Halosauridæ, 164
'Hassler,' American ship, 12
Herdman, on Ascidians, 143, 146
Hermit crabs, 139
Hoek, Dr., on Cirripedia, 132
— on Pycnogonida, 140
Holothuridea, 106
Hoyle, on Cephalopoda, 120
Hydroids, 92

'INVESTIGATOR,' H.M S., 16
Ipnops Murrayi, colour of, 60, 61

Ipnops Murrayi, phosphorescent organs of, 160
Isopoda, 127

KATANTIC sub-zone, 50
'Knight Errant,' H.M.S., 12

LAMELLIBRANCHIA, 119
'Lightning,' H.M.S., 8, 9
Lime, scarcity of, in bones of bathybial fish, 83
— in shells of mollusca, 83
Littoral sub-zone, 49
Lycodidæ, 155

MACRURA, 135
Macruridæ, 156
Madreporaria, 94
Medusæ, 91
Mollusca, 119
— colour of, 62
Moore, Capt., 94
Moseley, H. N., on colour of Cœlentera, 65
— on phosphorescent organs of *Ipnops*, 160
— on the darkness of the abyss, 22
— on the phosphorescence of Alcyonarians, 25
Murray, on *Bathypterois*, 159

NEKTON, 53
Neritic zone, 48
'Norma,' Mr. Hall's yacht, 12
Norske Nord-havns expedition, 7, 15

OPHIDIIDÆ, 156
Ostracoda, 123

PACKARD, on the illumination of the abyss, 23

Pediculati, 155
Pelagic zone, 47
Pennatulidæ, 96
Phoronis, 111
Phosphorescence of Alcyonarians, 81
— of deep-sea Crustacea, 80
— of Echinoderma, 81
Phosphorescent light in the abyss, 24
— organs of deep-sea fish, 77
Pigment in mucous membranes of deep-sea fish, 84, 152
Plankton, 52
Pleuronectidæ, 157
Polar currents, 30, 33
Polychæta, 118
' Porcupine,' H.M.S., 8, 9
Porifera, 91
Pourtales, Count, 10
Pressure in the abyss, 19
Protozoa, 88
Pteropod ooze, 39
Pycnogonida, 139

RADIOLARIA, 89
Radiolarian ooze, 39
Red mud, 37
— — off the Brazilian coasts, 42
Rhabdopleura, 111
Ross, Sir James, on the fauna of the deep sea, 3

SALMONIDÆ, 163
Sargasso sub-zone, 48
Sars, 6, 9,
— on Amphipoda, 125

Sars, on *Brisinga*, 105
Schizopoda, 133
Scopelidæ, 158
Serolis, 127
Siphonophora, 92
Size of deep-sea animals, 85
— of fish, 153
Smith, Mr., on Lamellibranchia, 119
Spatangoids, 101
Sponges, 91
Stebbing, Rev. T. R., on Amphipoda, 125
Sternoptychidæ, 158
Stomatopoda, 133
Stomiatidæ, 162

' TALISMAN,' French vessel, 12
Teleostei, 149, 154
Temperature of the abyss, 28
Thomson, Sir Wyville, on *Pourtalesia*, 10
— on *Thaumastocheles*, 137
— on the darkness of the abyss, 22
— on the phosphorescence of the sea, 26
Thoracostraca, 133
' Travailleur,' French vessel, 12
' Triton,' H.M.S., 12
Tunicata, 140

VEGETABLE life, absence of, 42
Verrill, on the illumination of the abyss, 23
' Vittor Pessani,' Italian vessel, 12
' Vöringin,' Norwegian vessel, 14

www.ingramcontent.com/pod-product-compliance
Lightning Source LLC
Chambersburg PA
CBHW020249170426
43202CB00008B/286